3000 800061 41647
St. Louis Community College

Meramec Library
St. Louis Community College
11333 Big Bend Blvd.
Kirkwood, MO 63122-5799
314-984-7797

WITHDRAWN

GLOBALIZATION AND HEALTH

INTERNATIONAL LIBRARY OF ETHICS, LAW, AND THE NEW MEDICINE

Founding Editors

DAVID C. THOMASMA†
DAVID N. WEISSTUB, *Université de Montréal, Canada*
THOMASINE KIMBROUGH KUSHNER, *University of California, Berkeley, U.S.A.*

Editor

DAVID N. WEISSTUB, *Université de Montréal, Canada*

Editorial Board

TERRY CARNEY, *University of Sydney, Australia*
MARCUS DÜWELL, *Utrecht University, Utrecht, the Netherlands*
SØREN HOLM, *University of Cardiff, Wales, United Kingdom*
GERRIT K. KIMSMA, *Vrije Universiteit, Amsterdam, the Netherlands*
DAVID NOVAK, *University of Toronto, Canada*
EDMUND D. PELLEGRINO, *Georgetown University, Washington D.C., U.S.A.*
DOM RENZO PEGORARO, *Fondazione Lanza and University of Padua, Italy*
DANIEL P. SULMASY, *Saint Vincent Catholic Medical Centers, New York, U.S.A.*
LAWRENCE TANCREDI, *New York University, New York, U.S.A.*

VOLUME 27

Globalization and Health

Challenges for health law and bioethics

Edited by

BELINDA BENNETT

*University of Sydney,
Australia*

and

GEORGE F. TOMOSSY

*Macquarie University,
Australia*

 Springer

St. Louis Community College
at Meramec
LIBRARY

A C.I.P. Catalogue record for this book is available from the Library of Congress.

ISBN-10 1-4020-4195-0 (HB)
ISBN-13 978-1-4020-4195-2 (HB)
ISBN-10 1-4020-4196-9 (e-book)
ISBN-13 978-1-4020-4196-9 (e-book)

Published by Springer,
P.O. Box 17, 3300 AA Dordrecht, The Netherlands.

www.springer.com

Printed on acid-free paper

All Rights Reserved
© 2006 Springer
No part of this work may be reproduced, stored in a retrieval system, or transmitted
in any form or by any means, electronic, mechanical, photocopying, microfilming, recording
or otherwise, without written permission from the Publisher, with the exception
of any material supplied specifically for the purpose of being entered
and executed on a computer system, for exclusive use by the purchaser of the work.

Printed in the Netherlands.

TABLE OF CONTENTS

CONTENTS

GLOBALIZATION AND HEALTH CARE

PREFACE

Within contemporary society, globalization has emerged as a key concern at the centre of ethical, legal and policy debates relating to health care. Conflicts between public interests and individual rights, the challenge of regulating health professionals and access to health services, and the effects of a global market all feature prominently in these discussions. As a result of globalization, these issues can no longer be understood solely within the political boundaries that define traditional notions of individuals and communities. Rather, solutions demand a global conception of rights and obligations, which in turn requires new approaches to health policy formulation and a reevaluation of existing ethical and legal frameworks. In essence, the impact of globalization on human health is testing the robustness of modern regulatory systems, legal doctrines and ethical paradigms.

PUBLIC HEALTH: DEVELOPING GLOBAL CONCERNS

The interconnectedness of the global economy presents new challenges in public health. While globalization has facilitated improvements in health care, it has also created new hazards and avenues for the exploitation of vulnerable persons. It is becoming increasingly apparent that both national and international responses are required. Indeed, as the chapters in this section convey, public health is rightly a global concern.

Globalization has led to a sharing of both risks and responsibilities in public health. Belinda Bennett reminds us of the ease with which infectious diseases can spread within the global community, given the speed of modern travel and trade. Despite a long history of the impact of infectious diseases on human society, the SARS crisis in 2003 demonstrated the ongoing importance of having efficient public health infrastructures at national *and* international levels. However, as Bennett notes, "the huge disparities in health and health infrastructure that exist between countries continue to undermine the ability of countries to respond rapidly and effectively to outbreaks of infectious disease."

Bennett's critique of the SARS crisis also shows how concerns about public health become acute at the interface between the developed and the developing worlds, which raises important questions about the meanings of rights and obligations in an international context. The evolving field of bioethics would play a vital role in addressing such dilemmas. Udo Schüklenk and Braimoh Bello argue, however, that much of bioethics discourse has focused on high-tech issues such as stem cell research and nanotechnologies, and "traditional liberal bread-and-butter issues of informed consent and individual autonomy." They argue that bioethics should instead focus on issues that affect many more people in the world – issues that address global inequities in health care between developed and developing countries. To that end, they propose a range of topics that should receive greater attention by professional bioethicists: the 10/90 gap in health research, the transnational organ trade, access to essential medicines, health-based immigration restrictions, international research ethics and the flow of health information. According to Schüklenk and Bello, a refocusing of bioethics as a field of inquiry is essential if it is to have continued contemporary relevance.

The evolution of public health as a global concern invokes questions about global social justice. As argued by George F. Tomossy and Joylon Ford, the quest for cures exposes fundamental deficiencies in legal doctrines insofar as they may prevent access to justice. They examine the plight of developing world subjects who may become injured in the course of first-world sponsored clinical trials, and who face significant legal obstacles when seeking compensation from multinational pharmaceutical corporations. Concerns about distributive justice thus come into conflict with the corporate incentive to pursue profits within a global market. Tomossy and Ford argue that citizens of one jurisdiction should not be exposed to risks of harm in order to benefit others, and would call upon investigators, sponsors and regulators alike to protect developing world subjects. They advocate that access to justice by developing world plaintiffs should be facilitated in first-world courts, which will require correcting procedural and substantive legal impediments that are presently almost insurmountable.

Finally, concerns about global social justice and public health invite consideration of the ethical grounds upon which arguments for obligations on the part of individuals, corporations and governments in the developed world towards developing countries might be based. This theme is explored by Deborah Zion, who analyzes obligations in terms of a duty of beneficence, efficacy, justice and integrity. She proposes that setting up processes to analyze the effects, burdens and benefits of clinical research would be a vital starting point towards relieving global health care inequities.

THE GLOBAL BIO-ECONOMY: CONSENSUS AND INNOVATION

The viability of national regulatory systems is continually being confronted by a global market for health care that is driven by the forces of innovation and health care consumerism. The emergence of a global bio-economy has created the need for transnational regulation of biotechnology and medical products. While generating consensus in health care policy formulation has always been a challenge, it is particularly so against the backdrop of globalization where consensus needs to be located at both national and international levels. And, as in the previous section, ethical issues permeate these discussions.

Derek Morgan argues that "we stand on the threshold of what might be thought to be a new dimension in the relationship of human sciences to biotechnology." He proposes that the emergent "bio-economy" is set to transform our lives in the same all-encompassing manner brought about by the industrial age and advent of the computer. With the key societal concerns for these stages having related to environmental degradation and privacy respectively, he predicts that the central issue in the new economy will be ethics. In order to resolve some of the current debates in this regard (for example, cloning, genetic patenting and bio-engineered foods), Morgan argues that the development of international consensus will require the implementation of "biomedical diplomacy," informed by traditional tools of "rhetoric, persuasion, negotiation, and economic and political leverage." The rationale for this process, he maintains, must be based on "rethinking equity in health," without which "all talk about human values, human dignity, human rights and democratic balance will be so much empty rhetoric."

Our understandings of "the global" and of "risk" help to shape responses to innovative technologies in health. Drawing upon the example of regulatory debates surrounding genetically modified foods in the United Kingdom, Alan Irwin considers the relationship between internationalized patterns of innovation and the development of national policy processes. His analysis reveals how differing conceptualizations of "the global" can exist within public discourses about innovation, and how the interaction between "the global" and "the national" affects the construction of regulatory debates. Irwin argues that these debates present political challenges in the need to formulate "more open cultures of deliberation and reflection," and that it is important "to move away from simply presenting globalization as an objective (and generally irresistible) force and towards an acknowledgement of its varied manifestations and social constructions."

Thomas Faunce explores the link between innovation and corporate globalization by examining the intersection of international trade law and domestic health policy. His critique addresses the impact of US-derived global intellectual property policies on government pricing of pharmaceuticals in Australia. He traces the evolution of these policies to their corporate origins in the United States and explores their enforcement through both international trade mechanisms and bilateral treaties. Faunce cautions that these policies represent "a significant, emerging problem for global public health," and urges greater attention to principles in bioethics, public health and international human rights in order to ensure affordable access to essential medicines.

As with trade, advertising is being recognized as a critical force in the global economy. Its relevance is particularly significant in today's consumer society where advertising plays a vital role in the development and expansion of markets for health products. Patricia Peppin analyzes the challenges associated with regulating advertising of pharmaceutical products through a comparative overview of the regulatory frameworks for advertising of medicines in the United States, Canada, the European Union, Australia and New Zealand. Drawing on semiotic theory, Peppin explores the construction of meanings through the information and images used in advertisements and the interpretation of those meanings by consumers. She warns of "significant public health consequences" associated with passing on advertising costs to health systems and with commodifying the doctor-patient relationship.

Globalization clearly presents significant difficulties for crafting consensus on regulatory policy in the area of biotechnology. This theme is explored by Timothy Caulfield and Barbara von Tigerstrom. Using the examples of gene patents and laws designed to limit human cloning, they reveal the competing tensions that emerge from global debates surrounding these issues. As they note, the demand for extensive regulatory intervention exists; however, "differing cultural and socio-political positions magnify the policy-making challenge." The authors acknowledge the difficulties inherent in reaching consensus on contentious issues and the potential for international agreements to limit the scope of national policy making. Their analysis thus yields an important lesson: "there can be no simple template for understanding and addressing the implications of globalization for biotechnology policy."

GLOBALIZATION AND HEALTH CARE

Having canvassed the implications of globalization for health care on a macro-level, this last section turns to the nexus between health care professionals and consumers. Globalization has had a fundamental effect on rights and obligations at the micro-level through its impact on national policies and legal systems. As these chapters show, the effects of globalization filter through to shape the rights of individuals and practices of the health professions.

Kerry Petersen's examination of the rights of children conceived using donated gametes to access identifying information about their biological (donor) parent provides a case for the study of individual rights in health care in a global setting. Despite the absence of consistent national or international patterns governing assisted reproductive technologies, common themes and regulatory approaches emerge from international comparisons. Petersen's critique thus reveals that incremental changes in regulatory reforms in this area favouring openness and disclosure of donor identity demonstrate the influence of human rights discourse on national health policy formulation.

John Harrington analyzes the impact of global market forces on national health systems, and signals the threat to national regulatory systems posed by health tourism and the commodification of human organs. Patients are increasingly travelling abroad in order to access health procedures. He argues that "consumption of health care, just like its provision, is no longer confined by national borders," with the global trade in human organs continuing to defy attempts to curtail it. Indeed, Harrington notes that the taboo against commodification has started to erode, with the consensus against commodification coming apart "under pressure of the actually-existing market."

In the final chapter in this collection, Ian Freckelton charts the emerging landscape in the global regulation of health care practitioners. Drawing on the experience of the United Kingdom, Canada, Australia and New Zealand, Freckelton maps the common regulatory trends that are emerging against the backdrop of this changing regulatory environment. He examines the changes caused by increasing consumerism and availability of health information in the age of the Internet, as well as the issues that arise from increased global movement of health professionals, and the ethical issues arising from the recruitment of developing world health professionals to meet the needs of health systems in developed countries.

CONCLUSION

Across the three themes of this volume, globalization has emerged as a fundamental force shaping ethical, legal and policy debates in health. The authors in this volume have shown that all aspects of health care, whether one is speaking of individual rights, professional obligations or governmental policy, are invariably influenced by transnational factors. As has been observed in globalization discourse more generally, these effects have been both positive and negative. The obvious challenge facing all countries, developing or developed, is to embrace the benefits of a global bio-economy while avoiding its harms. As is increasingly evident from attempts to govern innovation in biotechnology and access to health care, however, legal and regulatory mechanisms can only go so far towards achieving this goal. From the collective efforts of our colleagues in this volume, we would therefore derive the conclusion that a sound ethical base is needed upon which to ground policy initiatives, whether at national or international levels, and regardless of the difficulties obtaining political consensus might present. Such a base, we suggest, must ultimately be grounded in global considerations about equity and respect for human rights.

Belinda Bennett and George F. Tomossy
Sydney, September 2005.

ACKNOWLEDGMENTS

We wish to thank each of our esteemed authors who contributed their collective wisdom to this volume. Many of the chapters contained within this volume were presented in sessions at the *28th International Congress on Law and Mental Health,* held in Sydney, Australia in 2003, and subsequently revised for publication. In addition, we are honored to include original contributions from a number of colleagues who were unable to take part in the Sydney Congress. We also extend our gratitude to those who generously participated in the peer review process and provided authors with the benefit of their insights. We are likewise deeply indebted to Peter Hillerstrom and Roslyn Moloney for their assistance in preparing this book for publication and to Glenda Browne who prepared the index. Finally, we are grateful for the constant support and encouragement provided by Professor David N. Weisstub as Editor of the *International Library of Ethics, Law and the New Medicine,* and our colleagues at Springer, Fritz Schmuhl and Natalie Rieborn.

ACKNOWLEDGMENTS

CONTRIBUTORS

BELINDA BENNETT
Associate Professor, Faculty of Law, University of Sydney, Australia.

BRAIMOH BELLO
Epidemiologist, School of Public Health, University of the Witwatersrand, South Africa.

TIMOTHY CAULFIELD
Canada Research Chair in Health Law and Policy, University of Alberta, Canada.

THOMAS ALURED FAUNCE
Senior Lecturer, Australian National University Medical School and Law Faculty, Australia.

JOLYON FORD
Associate Member, Sydney Centre for International and Global Law, Faculty of Law, University of Sydney, Australia.

IAN FRECKELTON
Honorary Professor, Law School, Monash University, Australia.

JOHN A. HARRINGTON
Professor of Law, The University of Liverpool, UK.

ALAN IRWIN
Professor of Science and Technology Policy, The University of Liverpool, UK.

DEREK MORGAN
Professor of Health Law and Jurisprudence, Cardiff University, UK.

PATRICIA PEPPIN
Associate Professor, Faculty of Law and School of Medicine, Queen's University, Canada.

KERRY PETERSEN
Associate Professor, Faculty of Law and Management, La Trobe University, Australia.

UDO SCHÜKLENK
Chair in Ethics and Public Policy, Glasgow Caledonian University, Scotland.

GEORGE F. TOMOSSY
Lecturer, Division of Law, Macquarie University, Australia.

BARBARA VON TIGERSTROM
Assistant Professor, College of Law, University of Saskatchewan, Canada.

DEBORAH ZION
Medical Ethics Coordinator, Central and Eastern Clinical School, Faculty of Medicine, Nursing and Health Sciences, Monash University, Australia.

CHAPTER ONE

BELINDA BENNETT

TRAVEL IN A SMALL WORLD

SARS, globalization and public health laws

Late in 2002 reports of a life-threatening atypical pneumonia of unknown cause emerged in Guangdong Province in southern China. This disease, which we now know as Severe Acute Respiratory Syndrome (SARS), became the subject of a global alert by the World Health Organization (WHO) in March 2003, as reports of cases came in from China, Hong Kong and Vietnam (WHO 2003a).

Facilitated by international air travel, the spread of SARS quickly became a global phenomenon. By the end of July 2003, WHO had confirmed 8,098 cases of SARS, with 774 deaths across 29 countries (WHO 2003b). Clusters of cases were reported in China, Hong Kong, Taiwan, Singapore, Vietnam and Toronto in Canada (Tsang et al. 2003; Lee et al. 2003; Pang et al. 2003; Poutanen et al. 2003; WHO 2003b; 2003c). In an effort to stem the spread of SARS, WHO issued travel advice for some of the most affected areas (WHO 2003c, 10).

As the crisis unfolded, various guidelines were issued in an effort to halt the spread of the disease (e.g. WHO 2003d). Images of face masks, reports of quarantine and the growing number of infections and deaths from around the globe all contributed to a growing sense of panic. Yet with a concerted international effort the spread of SARS was limited and eventually halted. On 5 July 2003, WHO announced that human-to-human transmission of SARS had been interrupted (WHO 2003e).

This chapter reviews the impact of the SARS crisis and evaluates its significance in terms of global responses to the challenges posed by infectious diseases. The chapter also considers the role of the *International Health Regulations* in providing a legal framework for responses to disease outbreaks. Finally, this chapter will consider whether the SARS crisis advanced global preparedness for the next pandemic.

1

B. Bennett & G.F. Tomossy (eds.), Globalization and Health: Challenges for Health Law and Bioethics,
1-12.
© 2006 *Springer. Printed in the Netherlands.*

INFECTIOUS DISEASE IN THE GLOBAL VILLAGE

History provides us with some dramatic examples of the impact that a new infectious disease can have when it meets human populations with little or no immunity to the disease. In the Americas, Australia, Africa and the Pacific, indigenous populations were decimated in the years following the arrival of Europeans who brought with them germs and diseases such as smallpox, to which the local populations had no immunity (Diamond 1998, 210-4). Of course, European populations have not been immune to the ravages of infectious disease either. One-quarter of Europe's population was lost in the mid-fourteenth century to bubonic plague (id., 202), while the influenza epidemic at the end of the First World War killed an estimated twenty-one million people worldwide (ibid.). Furthermore, exposure to new tropical diseases, such as malaria and yellow fever, presented serious challenges to European attempts to colonize tropical areas (id., 214). Humanity has a long history, both in fighting infectious diseases and in searching for effective treatments or cures.[1]

Infectious diseases continue to reshape human society. Estimates from WHO indicate that in 2001 there were 14.7 million deaths caused by infectious diseases, a figure which accounted for 26 percent of total global mortality (Kindhauser 2003, 6). AIDS, tuberculosis and malaria combined accounted for 39 percent of deaths due to infectious diseases. These three diseases led to 5.6 million deaths in 2001. When the 5.8 million deaths from diarrhoeal disease and respiratory infections are also counted, five diseases caused 78 percent of the total infectious diseases burden (ibid.). Many of these diseases impact disproportionately on the world's poorest countries.

Seen in the context of these figures the SARS crisis may appear relatively unimportant. Yet SARS caught the world's attention precisely because of its potential to disrupt the populations and economies of the developed world, and can thus be seen in contrast to the diseases of the developing world:

> Everywhere, in the papers and on the internet, were images of commuters in Asia wearing surgical masks and empty airplanes and marketplaces. These images speak volumes, for in parts of the world where tuberculosis and AIDS and malaria reap their grim harvest of 6 million lives a year, there are not many trains, airplanes, or even masks.   (Farmer and Campos 2004, 247)

The significance of SARS is, also, its role as an indicator of the world's preparedness for the next global pandemic. Unlike other outbreaks of infectious disease that have eventually burned out, been geographically limited or susceptible to modern drugs or vaccines, the SARS epidemic was characterized by ready human-to-human transmission, the global spread of the disease (facilitated by international air travel), and the lack of any effective treatment or vaccine (Fidler 2004, 6). In this context then, "SARS posed a public health governance challenge the likes of which modern public health had not previously confronted" (ibid.).

New diseases continue to emerge while older diseases are increasingly resistant to the drugs used to treat them and threaten to reemerge as major health problems. In the latter decades of the twentieth century "over 30 new diseases – including AIDS and Ebola haemorrhagic fever – were detected for the first time" (Kindhauser 2003, 56). In 2004-5 outbreaks of avian flu in Asia have caused international concern about the potential for a new influenza pandemic (Barclay and Zambon 2004). Seen against this backdrop, SARS has been described as "the first severe and readily transmissible new disease to emerge in the 21st century" (WHO 2003c, 1). In this sense SARS may yet prove to be a great leveler, reminding us of the limits of our knowledge and ability to control microbes, and providing an impetus to even greater research on infectious diseases.

International trade and travel have long been mechanisms for the spread of infectious disease. In the era of globalization, the increase in international trade and travel has created an explosion of "microbial traffic" (Frenk and Gómez-Dantés 2002, 95), which has the potential to disrupt human health and society. In this age of global integration, outbreaks of infectious disease in one country can spread rapidly to other countries and even countries with sophisticated health infrastructures are not immune. As a Canadian report noted, "SARS has illustrated that we are constantly a short flight away from serious epidemics" (National Advisory Committee on SARS and Public Health 2003, 10). Being able to respond to outbreaks of infectious disease and assisting in capacity building of other countries to enable them to detect and respond to infectious disease is not only a global responsibility but "also a matter of enlightened self-interest" (ibid.).

International travel was clearly one of the contributing factors in the spread of SARS. In the days before air travel, international travel often meant weeks or months at sea. An outbreak of an infectious disease on board generally resulted in the ship being quarantined on its arrival in port.[2] With international air travel the time needed for an infectious disease to spread from one country to another is only as long as a trip on an airplane (Frenk and Gómez-Dantés 2002, 95). When we combine this potential with the number of airline journeys each year,[3] the difficulties in containing the spread of infectious diseases within the global village become apparent. The modern context for infectious diseases is evident from a comparison between two diseases: "It took smallpox centuries just to cross the Atlantic; a few weeks after arriving in Hong Kong from Guangdong, SARS had already spread to 30 countries on five continents" (National Advisory Committee on SARS and Public Health 2003, 197).

IMPACT ON HEALTH SYSTEMS

SARS provided a dramatic reminder, not only of the ability of diseases to spread internationally, but also of the need for health systems to have the capacity to respond to emerging crises with flexibility and efficiency. For developing countries in particular, new infectious diseases can overwhelm already burdened public health systems.

The speed with which SARS spread, its infective nature and the seriousness of the disease meant that the health systems in those areas that experienced clusters of cases were put under unusual strain (Cameron, Rainer and Smit 2003). This strain on health systems was exacerbated by the fact that in the early stages of the crisis many health workers themselves contracted the disease (ibid.; National Advisory Committee on SARS and Public Health 2003, 41; Pang et al. 2003, 3217). Health professionals were faced with the challenges of responding to increasing numbers of patients who were being struck down by a new mystery illness, as well as concern for colleagues who became ill and concern that their own families might become ill with the disease (Cameron 2003, 513; National Advisory Committee on SARS and Public Health 2003, 41).

In addition to striking health professionals, the SARS crisis also placed strain on other resources within the health infrastructure. Early symptoms of the disease are similar to those of other respiratory diseases, making early diagnosis difficult, which in turn placed further demands on health systems (National Advisory Committee on SARS and Public Health 2003, 1). Demand for surgical masks, gowns, gloves, thermometers and disinfectant all increased dramatically (Pang et al. 2003, 3217). In addition, in Beijing "76 new ambulances, 79 new radiograph machines, and 759 mechanical ventilators were acquired" and 123 fever clinics were set up at Beijing hospitals (ibid.).

The capacity to mobilize both health professionals and health resources rapidly in response to emerging health crises, whether from infectious disease or from other causes, is a vital element in the successful and rapid resolution of any health crisis. A report on Canadian responses to SARS indicated that the crisis "has reinforced the need for surge capacity" within the health system (National Advisory Committee on SARS and Public Health 2003, 102). The development of such capacity "is predicated on adequate professional resources, a depth of skill sets and overcoming jurisdictional legislative and regulatory barriers to allow, for instance, medical practitioners and health professionals to act outside their licensing jurisdiction in emergencies" (ibid.).

SARS AND THE GLOBAL ECONOMY

The SARS crisis highlighted both vulnerabilities and strengths in the global economy. The interaction of individuals, upon which so much domestic and international business depends, was reshaped at the height of the crisis. Within affected countries both official quarantine measures and general concern over the risk of infection served to limit the movement of people and impact upon domestic economies. SARS had the potential to impact on the global economy through declines in consumer demand, through reduced confidence "in the future of affected economies," and through increased costs related to disease prevention (Lee and McKibbin 2003, 4).

In Beijing all public entertainment venues were closed on 26 April 2003. "By the time these places began opening again during the second week in June, 3500 public places had been closed" (Pang et al. 2003, 3219). Schools closed on 24 April with some not reopening until July (ibid.). The economic impact of such disruption is substantial. According to a Canadian report on SARS, "Estimates based on volumes of business compared to usual seasonal activities suggest that tourism sustained a $350 million loss, airport activity reduction cost $220 million, and non-tourism retail sales were down by $380 million" (National Advisory Committee on SARS and Public Health 2003, 211).

However, it was not only tourism and travel-related companies that suffered. Businesses from unaffected areas that traded with affected areas also suffered. As fear of SARS kept people at home in affected areas, restaurants, shops and other businesses saw a decline in their business, a decline which in some cases also impacted on their suppliers from other unaffected areas (Bradsher 2003). During 2003, the fragile structure of the global economy was revealed as the SARS crisis hit harder.

PUBLIC HEALTH LAWS

Governments in countries affected by SARS were faced with the need to respond quickly so as to limit the spread of the disease within their populations. The screening of airline passengers for signs of illness and stories of quarantine were the images that captured media attention. Even in unaffected countries, governments responded by ensuring that public health laws were adequate to meet the challenge of SARS. In Australia, which was relatively unaffected by SARS, the federal government made SARS a quarantinable disease,[4] while state governments also amended their public health legislation to ensure that SARS was covered.[5]

With the reemergence of quarantine as a control measure in affected areas, the SARS crisis highlighted the tensions that can exist between public health and private rights (Mitka 2003). The history of quarantine tells the story of these tensions. With SARS and other new infectious diseases, quarantine is again being considered as a potentially important public health tool and the ethical issues raised by such measures are also receiving attention (ibid.). Surveillance, contact tracing, and travel restrictions, as well as isolation and quarantine, were important public health tools in the efforts to control the SARS outbreak and these measures also have implications for the rights of individuals (Gostin, Bayer and Fairchild 2003; Singer et al. 2003).

Despite the common use of these tools in SARS-affected countries, it is important to remember that "public health measures are embedded in broader sociopolitical contexts" of individual countries and reflect differing conceptions of the balance between the rights of individuals and governments (Gostin, Bayer and Fairchild 2003, 3231). While germs may not recognize borders, passports are, as Fidler (2004, 17) reminds us, "political phenomena," and "The politics of passports drive how human societies respond to the threats germs pose."

The SARS crisis of 2003 provides us then with an opportunity not only to reflect upon the efficacy of health measures and their success in limiting the spread of the disease, but also to reflect upon the legal and ethical issues that arise in the implementation of public health measures both domestically and in terms of their interface with the international community. Of course, in many respects it is easy to reflect upon these measures after the crisis has passed. There is, after all, perfect vision in hindsight. Yet, precisely because public health measures reflect the interface between the powers and duties of states and individuals (Gostin 2000), the manner in which public health measures are applied will make important statements about the value of individual rights within the community (Gostin, Bayer and Fairchild 2003).

Effective public health laws support the social structures and public policies that facilitate good health (Reynolds 1995, 1-2). Indeed law and the legal process have been described as "the inseparable companion of the public health process" (Reynolds 2004, 3). Effective public health laws also need to be able to respond to public health crises. In the wake of SARS and with heightened concerns over public health and national security, debates over public health and the relationship between individual rights and community interests have reemerged. While public health laws have traditionally reflected the tensions between public health and the liberty of individuals, during the latter part of the twentieth century these debates were recast and human rights and public health came to be seen as harmonious rather than in conflict (Childress and Bernheim 2003, 1196). As we reconsider the efficacy of our public health laws in the face of new twenty-first century challenges, we will also need to make key decisions about the nature of public health and the relationship between health and human rights (Gostin, Bayer and Fairchild 2003; Childress and Bernheim 2003).

GLOBAL PUBLIC HEALTH RESPONSES

The SARS story is one of both the successes and the failures of international public health. It is a story of international co-operation and collaborative scientific effort in the face of a major new global health challenge. However, it is also a story of the failure of existing international public health laws to respond adequately to the challenges posed by new infectious diseases such as SARS.

The potential for epidemics of infectious disease, the need for rapid and effective public health responses and the potential for disease outbreaks to have a serious impact on the economies of affected areas all lie at the heart of the challenge of developing effective global responses. Along with the globalization of public health (Yach and Bettcher 1998a; 1998b), we need the globalization of public health laws (Fidler 2002). It is in the interest of all members of the international community to ensure that our global networks and regulations are adequate to respond to emerging crises.

The speed of modern air travel adds a new dimension to the association between trade, travel and disease by allowing disease to spread more rapidly and more widely than was previously possible. This global nature of modern infectious disease has required global responses. In 2000, the World Health Organization brought together 112 existing networks into the Global Outbreak Alert and Response Network (GOARN) to maintain surveillance over infectious disease outbreaks (WHO 2003c, 4). GOARN has developed *Guiding Principles for International Outbreak Alert and*

Response, which operate with the aim of improving international co-ordination to support local efforts by partners in GOARN.[6] "From January 1998 through March 2002, WHO and its partners investigated 538 outbreaks of international concern in 132 countries" (WHO 2003c, 4). In addition, WHO has used the Global Public Health Intelligence Network (GPHIN) since 1997. GPHIN is a computer application which searches Internet sites for reports of disease (id., 4). There are significant benefits associated with GPHIN as it not only provides an early alert system for outbreaks of disease, but "also allows WHO to step in quickly to refute unsubstantiated rumors before they have a chance to cause social and economic disruption" (id., 5).

In response to the SARS outbreak, WHO sent teams of experts and equipment to countries requesting assistance. At the same time, based on the model of its influenza network, WHO established a virtual network of eleven leading laboratories through a shared website and teleconferences to work on identification of the cause of SARS and a reliable diagnostic test (WHO 2003c, 5). In the case of SARS, international scientific efforts led to the speedy identification of a coronavirus as the cause of SARS (Ksiazek et al. 2003; Drosten et al. 2003; Holmes 2003). Coronaviruses are not new to humans. Indeed up to 30 percent of human colds are caused by coronaviruses (Holmes 2003, 1949). However, it would appear that the coronavirus associated with SARS is new to humans. It probably came from a non-human host and developed the ability to infect humans (id., 1950). The link between diseases in animals and diseases in humans is not new. Many diseases that affect humans have evolved from diseases affecting animals or birds (Diamond 1998, ch 11). SARS is simply a recent example of this.

WHO also played an important role in the SARS crisis by issuing travel advisories for affected areas. The move to issue travel advisories was an important change for WHO since previously travel advisories had only been issued by individual countries (National Advisory Committee on SARS and Public Health 2003, 202; Fidler 2004, 137). Noting that "the effects of the travel advisories have been profound on the economies of targeted countries" (ibid.), a Canadian report on SARS concluded that "If WHO is to continue issuing advisories, clear criteria and a process for notice must be developed by agreement among member states" (id., 203).

The speed with which the international scientific community responded to the emergence of SARS indicates that globalization of the knowledge and information economies can provide positive outcomes for public health. Modern telecommunications, the use of the Internet, videoconferencing, webcasts and the international media all played important roles in disseminating information both to the scientific community and the general public and facilitated international scientific collaboration that led to rapid responses and collaborative research (Gerberding 2003; Drazen and Campion 2003). In 2003 WHO noted that "For continued progress against SARS, it is essential that we nurture the spirit of the unprecedented, global collaboration that rapidly discovered the novel virus and sequenced its genome" (WHO 2003f).[7]

THE INTERNATIONAL HEALTH REGULATIONS

The *International Health Regulations* (IHR) (WHO 1983) provide a framework for international public health law. In 1851 the first International Sanitary Conference was held in Paris following epidemics of cholera in Europe between 1830 and 1847. Between 1851 and the end of the nineteenth century there were ten conferences and eight conventions addressing the international spread of infectious diseases (WHO 2002, 1). Most of these international sanitary conventions did not come into effect, although conventions dealing with cholera and plague were adopted in 1892 and 1897 respectively (ibid.). In 1902, the International Sanitary Bureau was established and L'Office International d'Hygiène Publique was established in 1907 with a permanent secretariat in Paris (id., 2). The *International Sanitary Regulations* were adopted by WHO Member States in 1951. The Regulations were renamed the *International Health Regulations* in 1969 and, with modifications in 1973 and 1981, have been in force since (ibid.).

The Foreword to the IHR states that their purpose "is to ensure the maximum security against the international spread of diseases with a minimum interference with world traffic" (WHO 1983). Under the current IHR, WHO Member States are required to notify WHO of cases of yellow fever, plague and cholera. The IHR also set out requirements for health and vaccination certificates for travelers from infected to non-infected areas as well as health measures to be taken in relation to ships and aircraft and at ports and airports. The IHR set out the maximum measures that may be taken by WHO Member States during outbreaks of cholera, yellow fever and plague. In setting out the maximum measures that can be taken, the IHR provide a "template" for protective measures to ensure that other countries do not over-react and impose measures which are beyond those necessary from a public health perspective (WHO 2002, 3).

Unfortunately, the IHR have limited effectiveness. The current IHR rest on "an optimistic philosophy that infections can be stopped at borders by regulation of travellers, aircraft, and cargoes" (Nicoll et al. 2005, 322). The IHR focus on the spread of infectious diseases and do not address the prevention and control of infectious diseases by a state within its own borders (ibid; Fidler 2004, 33). Furthermore, since they only apply to three diseases – cholera, plague and yellow fever – the IHR do not apply to other infectious diseases that may also have serious implications for international public health. "Thus, the only international agreement on infectious diseases binding on WHO member states has been irrelevant to the SARS outbreak" (Fidler 2003a). Yet the current IHR have other shortcomings as well: they are dependent on affected countries making official notifications to WHO of cases of disease; there are few mechanisms in the IHR to foster collaboration between WHO and countries affected by diseases with the potential to spread internationally; the IHR lack mechanisms to encourage compliance by Member States; and WHO does not have the power to proscribe measures which will limit the international spread of disease (WHO 2002, 3).

Even before the outbreak of SARS, WHO had begun the process of revising the IHR. At its 56th annual meeting in 2003 the World Health Assembly adopted a resolution on the revision of the International Health Regulations in which the Assembly urged Member States *inter alia* "to give high priority to the work on the revision of the International Health Regulations and to provide resources and cooperation necessary to facilitate the progress of such work."[8] At the same meeting, the World Health Assembly adopted a resolution on SARS, which acknowledged, *inter alia*, "that the control of SARS requires intensive regional and global collaboration, effective strategies and additional resources at local, national, regional and international levels."[9] The Resolution urged Member States *inter alia*

> to commit fully to controlling SARS and other emerging and re-emerging infectious diseases, through political leadership, the provision of adequate resources, including through international cooperation, intensified multisectoral collaboration and public information.[10]

In May 2005 the World Health Assembly approved new *International Health Regulations*. These Regulations will come into force in 2007, two years after the date they were approved by the Assembly. Under the new *International Health Regulations,* Member States are required to assess public health events within their territory using an algorithm contained in the IHR. A Member State is required to notify WHO "of all events which may constitute a public health emergency of international concern within its territory" (Article 6(1)). Cases of certain listed diseases, including SARS and smallpox, must be notified to WHO. Other events must be assessed in terms of the seriousness of their public health impact, whether the event is unusual or unexpected, whether there is a significant risk of international spread, and whether there is a significant risk of international trade or travel restrictions.

The requirement under the revised IHR for notification of a "public health emergency of international concern" is an important development towards a new, more relevant IHR. The experience with SARS is also perhaps cause for cautious optimism. Although the IHR did not require countries to notify WHO of cases of SARS, "virtually all countries afflicted by SARS notified WHO of cases rapidly, continuously, and transparently" (Fidler 2004, 133).

Significant challenges remain. The huge disparities in health and health infrastructure that exist between countries continue to undermine the ability of countries to respond rapidly and effectively to outbreaks of infectious disease. In this sense then, the revised IHR must be simply a first step in a "process of health norm internalization and capacity building, bridging the gap between international goals and real-world public health preparedness" (Gostin 2004, 2627).

CONCLUSION

The SARS crisis of 2003 provided the international community with a wake-up call on the speed with which new infectious diseases can emerge, spread rapidly and develop into international public health crises. With the crisis having passed there is now the opportunity to reflect upon both domestic and international public health laws and to assess their adequacy for future health emergencies. The revision of the IHR and their new focus on public health emergencies of international concern is a important move away from specific diseases and towards a more flexible and responsive regulatory framework for infectious diseases. Yet this is only a small step towards global preparedness. Reducing the global burden of infectious diseases, particularly in developing countries, focusing on building public health infrastructure and capacity, and strengthening the frameworks for scientific collaboration and public health research are also vital steps towards an effective global response. Globalization does lead to an increased sharing of health risks in relation to infectious diseases. Only time will tell whether our domestic and international public health measures and public health laws will be adequate to respond to the next global health emergency in ways that are both effective and able to strike a balance between individual and community interests.

NOTES

An earlier version of this paper was presented at the 28th International Congress on Law and Mental Health in Sydney, Australia (2003). I am grateful to the conference participants for their comments and suggestions.

1. See for example, Edward Marriott's (2002) history of the search for the cause of plague and Fiammetta Rocco's (2003) history of quinine.
2. On the history of quarantine, see Fidler (2004, 27-9).
3. "According to World Tourism Organization data, approximately 715 million international tourist arrivals were registered at borders in 2002 (preliminary data)" (National Advisory Committee on SARS and Public Health 2003, 16).
4. *Quarantine Amendment Proclamation 2003 (No 1)* (Cth).
5. See e.g. *Public Health Amendment (Severe Acute Respiratory Syndrome) Regulation* 2003 (NSW).
6. World Health Organization, Global Outbreak Alert & Response Network, available at: <http://www.who.int/csr/outbreaknetwork/en/> (Last accessed: 24 January 2005).
7. For discussion of patenting and SARS see Rimmer (2004).
8. World Health Assembly, Resolution 56.28, para 3(1). Available at: <http://www.who.int/gb/ebwha/pdf_files/WHA56/ea56r28.pdf> (Last accessed: 4 May 2005).
9. World Health Assembly, Resolution 56.29, preamble. Available at: <http://www.who.int/gb/ebwha/pdf_files/WHA56/ea56r29.pdf> (Last accessed: 4 May 2005).
10. Id., para 1(1). For further discussion of this Resolution, see Fidler (2003b).

REFERENCES

Barclay, W.S., and M. Zambon. 2004. Pandemic risks from bird flu. *British Medical Journal* 328: 238-9.
Bradsher, K. 2003. The cough heard around the world. *Sydney Morning Herald* 25 April: 13.
Cameron, P.A. 2003. The plague within: An Australian doctor's experience of SARS in Hong Kong. *Medical Journal of Australia* 178: 512-3.
Cameron, P.A., T.H. Rainer, and P.D.V. Smit. 2003. The SARS epidemic: Lessons for Australia. *Medical Journal of Australia* 178: 478-9.
Childress, J.F., and R.G. Bernheim. 2003. Beyond the liberal and communitarian impasse: A framework and vision for public health. *Florida Law Review* 55: 1191-219.

Diamond, J. 1998. *Guns, germs and steel: A short history of everybody for the last 13,000 years.* London: Vintage.

Drazen, J.M., and E.W. Campion. 2003. SARS, the Internet and the *Journal. New England Journal of Medicine* 348: 2029.

Drosten, C., S. Günther, W. Preiser, S. van der Werf, H.R. Brodt, S. Becker, H. Rabenau, M. Panning, L. Kolesnikova, R.A. Fouchier, A. Berger, A.M. Burguière, J. Cinatl, M. Eickmann, N. Escriou, K. Grywna, S. Kramme, J.C. Manuguerra, S. Müller, V. Rickerts, M. Stürmer, S. Vieth, H.D. Klenk, A.D. Osterhaus, H. Schmitz, and H.W. Doerr. 2003. Identification of a novel coronavirus in patients with Severe Acute Respiratory Syndrome. *New England Journal of Medicine* 348: 1967-76.

Farmer, P., and N.G. Campos. 2004. New malaise: Bioethics and human rights in the global era. *Journal of Law, Medicine & Ethics* 32: 243-51.

Fidler, D.P. 2002. A globalized theory of public health law. *Journal of Law, Medicine & Ethics* 30: 150-61.

————. 2003a. SARS and international law. *ASIL Insights* April 2003. Available at: <http://www.asil.org/insights/insigh101.htm> (Last accessed: 24 January 2005).

————. 2003b. Developments involving SARS, international law, and infectious disease control at the Fifty-Sixth Meeting of the World Health Assembly. *ASIL Insights* June 2003. Available at: <http://www.asil.org/insights/insigh108.htm> (Last accessed: 24 January 2005).

————. 2004. *SARS, governance and the globalization of disease.* Hampshire and New York: Palgrave Macmillan.

Frenk, J., and O. Gómez-Dantés. 2002. Globalisation and the challenges to health systems. *British Medical Journal* 325: 95-7.

Gerberding, J.L. 2003. Faster…but fast enough? Responding to the epidemic of Severe Acute Respiratory Syndrome. *New England Journal of Medicine* 348: 2030-1.

Gostin, L.O. 2000. *Public health law: Power, duty and restraint.* Berkely and Los Angeles: University of California Press.

————. 2004. International infectious disease law: Revision of the World Health Organization's International Health Regulations. *JAMA* 291: 2623-7.

Gostin, L.O., R. Bayer, and A.L. Fairchild. 2003. Ethical and legal challenges posed by Severe Acute Respiratory Syndrome: Implications for the control of severe infectious disease threats. *JAMA* 290: 3229-37.

Holmes, K.V. 2003. SARS-associated coronavirus. *New England Journal of Medicine* 348: 1948-51.

Kindhauser, M.K. ed. 2003. *Communicable diseases 2002: Global defence against the infectious disease threat.* Geneva: World Health Organization (WHO/CDS/2003.15).

Ksiazek, T.G., D. Erdman, C.S. Goldsmith, S.R. Zaki, T. Peret, S. Emery, S. Tong, C. Urbani, J.A. Comer, W. Lim, P.E. Rollin, S.F. Dowell, A.E. Ling, C.D. Humphrey, W.J. Shieh, J. Guarner, C.D. Paddock, P. Rota, B. Fields, J. DeRisi, J.Y. Yang, N. Cox, J.M. Hughes, J.W. LeDuc, W.J. Bellini, L.J. Anderson, and the SARS Working Group. 2003. A novel coronavirus associated with Severe Acute Respiratory Syndrome. *New England Journal of Medicine* 348: 1953-66.

Lee, J.W., and W.J. McKibbin. 2003. *Globalization and disease: The case of SARS.* Working papers in trade and development, Division of Economics, Research School of Pacific and Asian Studies, Australian National University. August 2003. Working paper no. 2003/16. Available at: <http://rspas.anu.edu.au/economics/publish/papers/wp2003/wp-econ-2003-16.pdf> (Last accessed: 4 May 2005).

Lee, N., D. Hui, A. Wu, P. Chan, P. Cameron, G.M. Joynt, A. Ahuja, M.Y. Yung, C.B. Leung, K.F. To, S.F. Lui, C.C. Szeto, S. Chung, and J.J. Sung. 2003. A major outbreak of Severe Acute Respiratory Syndrome in Hong Kong. *New England Journal of Medicine* 348: 1986-94.

Marriott, E. 2002. *The plague race: A tale of fear, science and heroism.* London: Picador.

Mitka, M. 2003. SARS thrusts quarantine into the limelight. *JAMA* 290: 1696-8.

National Advisory Committee on SARS and Public Health. 2003. *Learning from SARS: Renewal of public health in Canada – A Report of the National Advisory Committee on SARS and Public Health.* Ottawa: Health Canada.

Nicoll, A., J. Jones, P. Aavitsland, and J. Giesecke. 2005. Proposed new International Health Regulations. *British Medical Journal* 330: 321-2.

Pang, X., Z. Zhu, F. Xu, J. Guo, X. Gong, D. Liu, Z. Liu, D.P. Chin, and D.R. Feikin. 2003. Evaluation of control measures implemented in the Severe Acute Respiratory Syndrome outbreak in Beijing, 2003. *JAMA* 290: 3215-21.

Poutanen, S.M., D.E. Low, B. Henry, S. Finkelstein, D. Rose, K. Green, R. Tellier, R. Draker, D. Adachi, M. Ayers, A.K. Chan, D.M. Skowronski, I. Salit, A.E. Simor, A.S. Slutsky, P.W. Doyle, M. Krajden, M. Petric, R.C. Brunham, A.J. McGeer, and the Canadian Severe Acute Respiratory Syndrome Study Team. 2003. Identification of Severe Acute Respiratory Syndrome in Canada. *New England Journal of Medicine* 348: 1995-2005.

Reynolds, C. 1995. *Public health law in Australia.* Sydney: Federation Press.

———. 2004. *Public health law and regulation.* Sydney: Federation Press.

Rimmer, M. 2004. The race to patent the SARS virus: The *TRIPS Agreement* and access to essential medicines. *Melbourne Journal of International Law* 5: 335-74.

Rocco, F. 2003. *The miraculous fever tree: Malaria, medicine and the cure that changed the world.* London: Harper Collins Publishers.

Singer, P.A., S.R. Benatar, M. Bernstein, A.S. Daar, B.M. Dickens, S.K. MacRae, R.E. Upshur, L. Wright, and R.Z. Shaul. 2003. Ethics and SARS: Lessons from Toronto. *British Medical Journal* 327: 1342-4.

Tsang, K.W., P.L. Ho, G.C. Ooi, W.K. Yee, T. Wang, M. Chan-Yeung, W.K. Lam, W.H. Seto, L.Y. Yam, T.M. Cheung, P.C. Wong, B. Lam, M.S. Ip, J. Chan, K.Y. Yuen, and K.N. Lai. 2003. A cluster of cases of Severe Acute Respiratory Syndrome in Hong Kong. *New England Journal of Medicine* 348: 1977-85.

WHO (World Health Organization). 1983. *International Health Regulations.* Geneva: World Health Organization. Available at: <http://www.who.int/csr/ihr/en/> (Last accessed: 4 May 2005).

———. International Health Regulations Revision Project. 2002. *Global crises – Global solutions: Managing public health emergencies of international concern through the revised International Health Regulations.* Geneva: World Health Organization (WHO/CDS/CSR/GAR/2002.4).

———. 2003a. *WHO issues a global alert about cases of atypical pneumonia: Cases of severe acute respiratory illness may spread to hospital staff* (Media Release, 12 March 2003).

———. 2003b. *Summary of probable SARS cases with onset of illness from 1 November 2002 to 31 July 2003.*

———. 2003c. *Severe Acute Respiratory Syndrome (SARS): Status of the outbreak and lessons for the immediate future.* Geneva: World Health Organization, 20 May.

———. 2003d. *WHO recommendations on SARS and blood safety.* 15 May. Available at: <http://www.who.int/csr/sars/guidelines/bloodsafety/en/print.html> (Last accessed: 24 January 2005).

———. 2003e. *Update 96 – Taiwan, China: SARS transmission interrupted in last outbreak area.* 5 July.

———. 2003f. *Patent applications for SARS virus and genes.* 29 May.

Yach, D., and D. Bettcher. 1998a. The globalization of public health, I: Threats and opportunities. *American Journal of Public Health* 88: 735-8.

———. 1998b. The globalization of public health, II: The convergence of self-interest and altruism. *American Journal of Public Health* 88: 738-41.

CHAPTER TWO

UDO SCHÜKLENK & BRAIMOH BELLO

GLOBALIZATION AND HEALTH

A developing world perspective on ethical and policy issues

Globalization in the context of health may be viewed as a concerted global effort towards prevention, elimination and eradication of diseases, and the promotion of human health worldwide. This brings a lot of issues to mind: from the ever-growing mobility of people and the public health consequences of such a change in peoples' movement patterns, to concerns about ethical standards in international health research, as well as ethical analyses critiquing the continuing existence of the 10/90 gap in health research. Intellectual property rights and affordable access of the world's poor to patent-protected essential AIDS drugs, as well as immigration ethics, have been high on international agendas. What is significant, perhaps, is that most of the critical contributions to any of these ongoing concerns have not actually come from professional bioethicists but from other professionals (Farmer and Gastineau 2004). Bioethicists, true to the traditions of their relatively young field, have continued their focus on high-tech issues such as stem cell research and nanotechnologies (Dhai et al. 2004; Boyle and Savulescu 2001; Daar et al. 2004). Some bioethicists have criticized this preoccupation with "high-tech" issues as "following the money" (Elliot 2004).

The argument advanced here is, basically, that funding for bioethics research is to a large extent controlled by the pharmaceutical industry and government organizations, such as the US National Institutes of Health, or wealthy private funders, such as the Wellcome Trust in the UK. Because all of those funders are mainly concerned about "hi-tech" issues as opposed to developing world issues, not much bioethical research is undertaken in those areas. Considering that many more lives are affected by the existence of intellectual property rights and the frequent resultant lack of affordable access to essential drugs, this in itself is worthy of critical reflection about the field and its researchers' priorities. This contribution aims to sketch an overview of issues that should take priority within bioethics over the traditional liberal bread-and-butter issues of informed consent and individual autonomy. By necessity, such an overview cannot at the same time provide in-depth

13

B. Bennett & G.F. Tomossy (eds.), Globalization and Health: Challenges for Health Law and Bioethics, 13-25.
© 2006 *Springer. Printed in the Netherlands.*

analyses of each of the issues mentioned. Rather, it should be read as a proposal for a different kind of research agenda for bioethics to what is the field's current status quo.

THE 10/90 GAP

According to the Global Forum for Health Research (2002, 3):

> Every year, more than US$70 billion is spent on health research and development by the public and private sectors. An estimated 10% of this is used for research into 90% of the world's health problems. This is what is called the 10/90 gap.

The issue of neglected diseases research is obviously of central importance to people living in developing countries (Médecins Sans Frontières 2001). However, even if the current commercial research agendas of the pharmaceutical industry were directed at diseases according to the world's needs and not necessarily according to commercial considerations, another problem would inevitably arise. It was brought to the world's attention by Edwin Cameron, an HIV-positive South African judge. In his address to the delegates attending the International AIDS Conference in Durban in 2000 he stated:

> Nearly 34 million people in our world are at this moment dying [of AIDS]. And they are dying because they don't have the privilege that I have, of purchasing my health and life...Now why should I have the privilege of purchasing my life and health when 34 million people in the resource poor world are falling ill, feeling sick to death, and are dying? That to me...seems a moral inequity of such fundamental proportions that no one can look at it and fail to be spurred to thought and action about it. That is something, which we in Africa cannot accept. It is something that the developed world also cannot accept.   (Cameron 2000)

In other words, people in developing countries will face two major challenges in their attempts to access lifesaving medication. The first challenge is that the medication in question might not exist, because the necessary research was not considered commercially sufficiently attractive by pharmaceutical multinationals to justify substantial investment (Schüklenk and Ashcroft 2002; Sterckx 2004). The second challenge is that the medication available is likely to be too expensive to permit affordable access. This has been the case in many African countries like South Africa and Uganda, where the majority of people suffering from HIV/AIDS have no access to treatment drugs because of their extreme cost. In Uganda, for example, where more than two million people are infected with the virus, it is estimated by UNAIDS that not more than 2,000 people have access to drugs (Schüklenk and Ashcroft 2002; Cochrane 2000). A team of researchers from Doctors Without Borders (Médecins Sans Frontières) reviewed the situation in a 2002 *Lancet* publication and concluded that "despite an ever-increasing need for safe, effective, and affordable medicines for the treatment of these [tropical] diseases, drug development has virtually stopped" (Trouiller et al. 2002).

International discussions about how to address these challenges in a just manner have been going on for some time but without any satisfactory solutions at the time of writing (Hubbard and Love 2004). New trade frameworks for health care research and development (R&D) have, for instance, been proposed by the US-based Consumer Project on Technology.[1] Love (2003, 1) argues that "trade agreements should be reframed to focus on standards for sharing the costs of R&D" and that we "need business models for financing R&D that do not depend on marketing monopolies for approved products." An ethical allocation of research resources that takes into account global health needs would help to bridge this gap and reduce the consequences of such misallocation. This would not only be of benefit to the researchers involved, but also to their countries and the global community in general.

ORGAN "DONATION"

Unprecedented numbers of people move across borders (sometimes illegally), but so do human organs destined for organ transplantation. Ethical issues discussed in this context have to do with the fundamental concern of exploitation of impoverished organ donors in the developing world. Commercial organ sales resulting in organ transfers (for payment) from developing world sources to developed world recipients clearly are of concern, because of the economic power differential between the parties to the contract. Similarly, doctors' participation in such contracts has been criticized as incompatible with their professional obligations.

Professionalism, in its historic religious derivation, means essentially to profess publicly to serve the public good (Koehn 1994). Arguably, access to lifesaving organs based exclusively on a given prospective patients' capacity to pay is incompatible with professional duties so understood. Equitable access as a fundamental requirement of the principle of distributive justice is not guaranteed under such circumstances, despite frequently equal needs on the part of those requiring transplant organs. The material principle of justice requires the equal treatment of equal needs. Medical professionalism traditionally required impartiality in the treatment of patients competing for scarce resources. If recipients of donor organs are allowed to jump the queue based on their capacity to pay, as opposed to being prioritised based on clinical need, medical professionalism ceases to operate as it should.

Evidence has begun to accumulate that at least some "donors" had no substantial understanding of the implications of their decision, or were not even aware of their "donation." It has also been reported that organs have been stolen from deceased patients as well as from impoverished patients. The latter had gone to hospitals for care and left them with less than their full set of organs (*Associated Press* 1999). Reportedly, the Turkish Interior Ministry warned after the major 1999 earthquake that groups were roaming the country in search of children orphaned by the earthquake to kidnap them and remove organs for sale (*Bloomberg* 1999). Many reports have also been accumulated by University of California at Berkeley-based Professor of Anthropology, Nancy Scheper-Hughes, who noted that:

The rapid transfer of organ transplant technologies to countries in the East (China, Taiwan, and India) and the South (especially Argentina, Chile, and Brazil) has created a global scarcity of viable organs that has initiated a movement of sick bodies in one direction and of healthy organs – transported by commercial airlines in ordinary Styrofoam picnic coolers conveniently stored in overhead luggage compartments – often in the reverse direction, creating a kind of "kula ring" of bodies and body parts.

(Scheper-Hughes 2000, 193)

IMMIGRATION ISSUES

People from developing countries face ever-growing hurdles in their attempts to migrate to developed countries. A variety of means have been introduced to prevent sick migrants from entering developed countries. For instance, in California, regulations require that health care providers who rely on "public funds" not care for illegal immigrants. In Canada, the identification of potential immigrants who might cause costs to taxpayers has become one of the main purposes of the country's immigration medical procedures. In the United Kingdom, prospective immigrants have to undergo a medical examination designed to determine whether they might be unable to support themselves or their dependents or require medical treatment.

The United Kingdom Home Office can grant immigrants permanent leave of stay for compassionate reasons. One would expect that a serious, even life-threatening illness of a person from a developing country should be sufficient to grant permanent leave of stay on compassionate grounds. Yet the opposite is the case. Medical doctors can be called in to prevent immigration. Among the reasons they could offer to justify the refusal are contagious diseases such as pulmonary tuberculosis, but also senility, mental disorders, sexual aberration, physical disability (including bodily deformity), and other conditions that might prevent persons from supporting themselves or their dependents. In other words, vulnerable people from countries that can reasonably be assumed to have insufficient health care facilities are most likely to be denied entry into developed countries and denied access to their health care facilities.

Medical professionals across the Western world are continuously involved in medical examinations designed to prevent the most vulnerable of potential immigrants from entering societies with fully functional medical systems. Indeed, doctors within the public health care system have been encouraged to report sick, illegal immigrants to the authorities to help them deport such people. Recently, Pollard and Savulescu (2004, 349) proposed changing the eligibility criteria for access to National Health Service treatment to "urgent medical treatment" by overseas visitors (to the United Kingdom) and people of uncertain residential status. They provide the following definition of "urgent medical treatment": "treatment to prevent either risk to the public (in the case of infectious disease) or serious harm to the individual (now or in the future)."

An international organization of medical doctors, Physicians for Human Rights, has noted that fundamental challenges to the notion of medical professionalism are posed by what it describes as "dual loyalties" conflicts. These conflicts occur whenever doctors find themselves in a situation where they have to decide whether they should prioritize the health of their patients over government regulations, policies of managed care organizations and the like (Physicians for Human Rights 2002). The obvious question is whether health care professionals have moral obligations to their patients that preclude participation in such practices. Unsurprisingly, immigrants' organizations have suggested that doctors should have the health of their patients in mind, and support their attempts to immigrate into a country that has a functioning health care system. At least some immigrants have been reported to immigration authorities after having sought medical assistance. It has been argued in the *British Medical Journal* by leading UK bioethicists that it is up to immigration authorities, not doctors, to enforce immigration policies (Pollard and Savulescu 2004).

The participation of medical professionals in immigration control measures is ethically problematic and deserves closer scrutiny by medical ethicists as well as national medical associations. In addition, an inquiry is called for into the professional ethics of doctors who join forces with immigration authorities in order to track down illegal migrants. Clearly, this issue needs to be situated within the context of distributive justice. After all, it is Western pharmaceutical companies that charge prices for their drugs that make lifesaving medications effectively inaccessible to people in the developing world. Is it so unreasonable, then, to suggest that there is a moral obligation to support sick migrants from such societies who seek medical support in the developed world?

IPR AND ACCESS TO ESSENTIAL DRUGS

Bioethicists and political philosophers have joined in an international debate centering on the relationship of intellectual property rights (IPR), patent protection, drug prices and access to essential drugs for the poor in developing countries. At issue is the fact that pharmaceutical multinationals charge prices for patented drugs that prevent the overwhelming majority of people in developing countries from access to their products (Sterckx 2004). The primary reason for the protection of intellectual property rights and patents is, interestingly, a public interest-based argument. It states that in order for such companies to be able to achieve a desirable public objective, that is to research and produce new drugs, they must be able to reap or anticipate a profit from their products sufficiently high to support a significant research effort (Schull 2000). Precisely this reason, however, provides a possible justification for overriding some patents in some countries. If patents are protected to ensure that the public has continuing access to privately developed drugs, something is going wrong when the public cannot actually afford access to those drugs at all. This is not to say that private companies that spend enormous amounts of money on research should not recoup their investment from patented drugs. However, if the owners of those patents price their drugs out of the reach of the majority of those in need, so goes the argument, there is very little point in protecting those patents any longer (Schüklenk and Ashcroft 2002).

Of course, the issues are more complicated than this. For a start, in developed countries, many people in need of expensive, patented AIDS drugs are able to access those drugs. The same, however, cannot be said for developing countries. Here the argument in favor of setting aside patents seems to succeed. Indeed, the World Trade Organization's *Trade-Related Aspects of Intellectual Property Agreement* (TRIPS) expressly permits developing countries to override the patents for drugs that could be utilized in public health emergencies such as AIDS, TB and malaria. The problem with TRIPS-related policies is, as argued by the pharmaceutical companies, that without such internationally recognized patent rights, there would be no funds and incentives for further R&D (Schull 2000).

Currently, patent owners already battle for market share for their existing products. The argument is that financial incentives to engage in R&D investment may not necessarily jeopardize drug research, as they do not tamper with the patent rights in place in the developed world. Furthermore, it is worth noting that the profits for which industry is clamoring have never been generated in the developing world. What is interesting with regard to this debate is that bioethicists have moved on from arguments suggesting that pharmaceutical multinationals should provide some donations to worthy projects, and engaged in argument at a policy level. The discussion, in other words, is no longer operating on the usual ethics level of "please do the right thing," and investigating "what is the right thing to do," rather it has moved on to proactive policy development (Cohen and Illingworth 2003). It certainly helps that legal frameworks exist to actually execute this strategy. Of course, bioethical analyses mirror in many ways the ideologies commonly found in public discourse. Not surprisingly, neo-liberal philosophers have argued in favor of patent protection, and suggested that developing countries do more to support important drug research on their soil (Resnik 2001).

It is worth asking the following questions: who should invest in R&D in poorer countries? How feasible will it be to generate R&D funding outside pharmaceutical companies? If this was feasible, drug patents may begin to fade away from the center of international ethical debate. It is already being argued that

> taxpayers, not shareholders, have borne most of the cost. Publicly funded research organizations have contributed hundreds of millions of taxpayers' dollars to AIDS drug research. Indeed, the Pharmaceutical Research and Manufacturers of America, an industry lobby group, estimates that private industry finances only about 43% of drug development. Five commonly used drugs against AIDS – didanosine, lamivudine, nevirapine, stavudine, and zidovudine – were developed largely as a result of public funds.   (Schull 2000)

As Tina Piper (2004, 475) argues:

> There is a growing recognition in the international and domestic context that the patent system may not function well for the equitable and affordable delivery of healthcare goods, and its strict operation may have to be altered by enacting accompanying regulation or changing patent laws themselves.

RESEARCH ETHICS

Most major pharmaceutical companies have transformed into global corporations. Their research no longer takes place exclusively in developed countries. An ever-growing number of clinical trials are undertaken in developing countries capable of hosting such investigations (Concar 2004, 28). On the positive side, this globalization of research activities results in the desirable strengthening of research capacities in developing countries. It is also undoubtedly cheaper to run some trials in developing countries than it would be to run them in a developed country. This certainly is one of the reasons why research has become a global enterprise. To a large extent this is what globalization is all about. Producing certain outcomes or products as cost effectively as is feasible remains the primary objective. Achieving this objective is conceptualized and operationalized not any longer in a particular country but on a global scale.

Concerns about the exploitation of research participants in developing countries focus on two distinct issues: the standards of clinical care provided during such trials and the question of what care or benefits are ethically required after the trial has concluded.

An acrimonious and lengthy international debate on the ethically appropriate standards of care provided in clinical trials in developing countries was triggered by an AIDS trial that took place in South Africa, among other countries. At the heart of this debate is whether the standard of care is universal or local (determined by the levels of a given country's infrastructure and wealth).

The trial in question was responding to a perceived need in developing countries to investigate a potentially affordable medication that might reduce the transmission of HIV from a pregnant infected woman to her offspring. Before the trial started, a trial undertaken only in developed countries had established that 25 percent of HIV-infected pregnant women who use no antiretroviral medication transmit the virus to their offspring, but less than 8 percent of those who used a particular regime of the drug zidovudine do. This trial regime soon became the gold standard for mother-to-child-transmission (MTCT) prevention in the developed world. Unfortunately, this drug regime proved to be unaffordable for the majority of infected women in developing countries. The developing world trial aimed at testing a cheaper yet effective MTCT regime. In order to do so, the investigators sought to establish not whether a lower dosage was as efficient as the higher one provided in the developed world, but whether it was more efficient than doing nothing. In other words, the test regime was matched against a placebo control.

The trial regime was quickly denounced in international medical as well as bioethics journals, because it violated a standard provision of the world's pivotal research ethics guideline, the World Medical Association's *Declaration of Helsinki* (Lurie and Wolfe 1997; Schüklenk and Ashcroft 2000). The Helsinki Declaration would have required the developing world investigators to compare their trial regime against the existing gold standard of HIV MTCT prevention, i.e. the locally unaffordable regime.

Defenders of the trial in question argued that the Helsinki Declaration was not a particularly useful ethical yardstick to hold against the developing world-based trial. They proposed that the Declaration be adjusted to reflect the realities of developing world health care (Levine 1999). At the heart of their analysis were two ethical arguments. The first argument was that women in the trials in question were no worse off by virtue of their trial participation. The participating pregnant women would not have had access to antiretrovirals at all, in the public health care sector of their countries. Hence, those in the placebo arm were not in any way worse off, while those in the active agent arm were almost certainly better off. The second argument states, in a nutshell, that the Helsinki requirement to test a new experimental agent always against the best-proven diagnostic and therapeutic method of treatment is preventing useful research in developing countries. If one always has to test a new agent against the best medication that exists somewhere in the world, no matter whether people in a given location can access or afford it, one would not be able to investigate cheaper treatment regimes that could actually make a difference in a developing country.

The counter arguments employed questioned the need to develop cheaper drugs, asking whether or not there is such a thing as a natural (in the sense of: not subjected to human choices) price for any given medication. Proponents of this point of view argue that the prices people pay for essential drugs are the result of pricing decisions made by major multinational pharmaceutical corporations (Schüklenk 2000). Solutions to the matter of drug pricing then could be sought in internationally accepted regulations such as the TRIPS Agreement, which allows for compulsory licensing and parallel importation of essential drugs in cases of public health emergencies. This obviously would eliminate a lot of the persuasive power of the position we have just described. If it is possible to remove the patent protection of essential AIDS drugs, and as a corollary of this their high price tag, the need to develop cheaper AIDS drugs for developing countries may well all but disappear. Arguably, however, this does not in itself solve other issues such as inadequate systems for the delivery of antiretroviral drugs in many developing countries. The successful provision of such treatment regimes relies on regular counts of certain blood cells, satisfactory levels of nutrition and so on and so forth. Still, it is widely accepted that the drug price is one of the primary hindrances of poor peoples' access to essential AIDS drugs. One might even argue that prices for diagnostic instruments could be subjected to a critique similar to that deployed against high prices for patented drugs. The point of this argument being is that diagnostic instruments' prices are also kept artificially high because of the patents held by the developer and producers of such products.

Two other interrelated issues need to be looked at. Both have to do with standards of care. One such issue is what standards of care participants can reasonably expect after a trial has concluded; the other question is what standards of care are ethically required for people who contract HIV during preventative vaccine or microbicide trials? Preventative vaccine or microbicide trials are designed to test whether a vaccine candidate or a microbicide is capable of reducing the number of

HIV infections transmitted in a given cohort of patients. As no preventative agent currently exists it is legitimate to compare the candidate agents against a placebo control. Invariably, in any given clinical trial, there are a number of participants who labor under a therapeutic misconception. They believe, despite the genuinely best efforts of the investigators, that they are receiving a working preventative agent against HIV. If such a participant becomes infected as a consequence of risk-taking triggered or influenced by the therapeutic misconception, such a participant's situation has worsened as a consequence of his or her trial participation. UNAIDS holds the view that an infection acquired under such circumstances does not constitute a trial-related injury. While it might be politically and financially opportune, we doubt that this view withstands critical scrutiny (Schüklenk 2004).

The policy implications of this question obviously go far beyond AIDS research and affect all prevention trials. In South Africa two pivotal research ethics committees withdrew approval for upcoming preventative HIV vaccine trials because the committees considered the investigators' intention not to provide essential AIDS drugs to trial participants who become infected unethical (Schüklenk 2004). The ethical evaluation of this question is once again strongly influenced by economic factors. If a given investigator has to find the additional funds to guarantee life-long treatment (or, as South African ethics committees suggested, while there is a therapeutic benefit to be gained) with essential AIDS drugs, he or she may well decide not to proceed with the trial at all. In turn, this could severely increase the time it takes to develop a preventative HIV vaccine. More lives might be lost in the medium to long-term if this was a consequence of demands pertaining to standards of care to be provided to failures in preventative trials.

FLOW OF HEALTH INFORMATION

The concept of globalization is embedded in the speedy development of technologies that have enhanced communication and interconnectedness, and have led to the rise of claims that the world has become a global village. Though it is often said to bring new opportunities for sharing ideas and technologies, it is uncertain how far this has been realized in terms of global health. What is presently apparent is that even though these technologies have been around for decades, their potential to transform the health of the global community through a fluid health information system has not been harnessed. The resource constraints in developing countries do not promote the development or acquisition of these technologies. However, information is one of the pillars on which an efficient health information system is built. This has severely hampered capacity development in these regions. As at 1998, less than 0.2 percent (one million people of the 700 million people in Africa) has access to the Internet, and 80 percent of these were in South Africa. Of the remaining 20 percent, the ratio of Internet subscribers to those who are not is 1:5000, a figure that is quite alarming when compared to the 1:5 ratio obtainable in the US or Europe (Edejer 2000). Recent statistics show that the digital divide is still very severe: as at 2004, only 1.4 percent of Africans and 7.1 percent of Asians have access to the Internet compared to 68.3 percent in America and 45.1 percent in the European Union. In Nigeria and Ethiopia, only 0.5 percent and 0.1 percent respectively use the Internet as compared to the 75 percent and 69 percent for Sweden and America respectively (Internet World Statistics 2005). As Edejer (2000, 798) observes: "That the digital divide is more dramatic than any other inequities in health or income is depressing because information and communication technologies have been hailed as one of the potential solutions to these inequities."

Efforts are being made to reduce if not eliminate these inequalities, but as more people have access to Internet-delivered information in the developing world another problem (that really constitutes an extension of the IPR discussion above) lies in the fact that vital health information needs subscriptions in order to be accessed. Subscription fees charged by some publishers are too high for the average researcher, and frequently even institutions in developing countries cannot afford them. A look at the references of the published work of young scholars reveals that the bulk of them have come from freely accessed journals, some of which are beginning to cancel their current free-access policy. The *British Medical Journal* will in 2005 revert to a subscribers-only access policy plus free online access for readers in least developed countries. Ironically readers in least developed countries are also least likely to have Internet access. South Africa, for example, is quite rightly not listed among this group of "least developed countries"; hence, free online access to the *British Medical Journal* will cease to exist. The predicament for such countries is that as they get their infrastructure sufficiently developed to acquire the relevant information and to utilize it, they are denied access to valuable information because they are not considered poor enough any longer to be given the concession of a free subscription. Yet they are not rich enough to continually afford the subscription fees and/or they must choose which of a range of essential journals will be provided.

Other journals have proposed a different model: they offer free subscription but require that authors pay for their work to be published. This makes it difficult for young researchers to publish their research findings, and is therefore disrupting the flow of vital information that such journals are supposed to serve. This lack of equitable access to health information is undoubtedly one of the reasons why very few works from researchers in developing countries are published in internationally recognized journals. Of about 3,000 journals indexed by international services like Medline and Science Citation, only 2 percent are from the developing world while the other 98 percent come from the developed countries. The reasons for this are not so clear. According to Edejer (2000, 798):

> possible explanations range from the difficulties encountered by researchers in developing countries in gaining funding for research – only 10% of funding is spent studying problems relevant to developing countries – to the existence of ethnocentrism at its worst in biomedical publishing circles.

Although, global health problems due to inequalities in health information flow are plenty, some movement towards more equitable access to health information could be achieved through concerted efforts from developed and developing world leaders. The latter have, through bad governance and corruption, plunged many developing nations into abject poverty. Developed countries, which are the main custodians of novel ideas and technologies, should formulate information dissemination policies in the spirit of globalization as this will improve global public health.

CONCLUSION

Globalization in its own right is not necessarily a bad development for both developed countries and the developing world as it has helped to break down barriers between regions, eliminate monopolies, increase markets for pharmaceutical companies and promote access to drugs and health care for developing countries. The international flow of research funds has undoubtedly assisted in building up research infrastructure and competencies in developing countries. Global efforts have been largely successful in eradicating smallpox, eliminating polio, which is likely to be eradicated in the near future, and controling many other diseases like onchocerciasis and tuberculosis (Global Forum for Health Research 2002).

Readers, we hope, will agree with us, that professional bioethics could benefit from efforts to refocus its research agenda by way of taking into account the number of people affected by a particular concern. Our overview, as indicated in our introductory remarks, makes no pretense to analyze the issues and answer the problems mentioned in this article. Rather we suggest that the mentioned areas of inquiry are worthy of serious bioethical analysis and arguably that they are of greater significance than the areas that are preoccupying bioethicists' research time at the time of writing. We further suggest that bioethics cannot avoid addressing economic inequalities as well as international policies if it wishes to have anything of relevance to contribute to these issues. No doubt, this would have negative implications for bioethics' funding, seeing that major funders of bioethical research are not interested in offering resources to undertake this necessary work. Arguably the integrity of the field itself could be at stake, depending on how mainstream bioethics responds to this challenge.

NOTES

1.   See the wealth of documents at Consumer Project on Technology, available at: <http://www.cptech.org> (Last accessed: 24 January 2005).

REFERENCES

Associated Press. 1999. Officials probe Turkish organ theft, 17 September.
Bloomberg. 1999. Turkish officials say quake children stolen for their organs, 1 September.
Boyle, R.J., and J. Savulescu. 2001. Ethics of using preimplantation genetic diagnosis to select a stem cell donor for an existing person. *British Medical Journal* 323: 1240-3.
Cameron, E. 2000. Press Conference, XII International AIDS Conference, Durban, South Africa, 10 July.

24   SCHÜKLENK & BELLO

Cochrane, J. 2000. Narrowing the gap: Access to HIV treatments in developing countries. A pharmaceutical company's perspective. *Journal of Medical Ethics* 26: 47-50.
Cohen, J.C., and P. Illingworth. 2003. The dilemma of intellectual property rights for pharmaceuticals: The tension between ensuring access of the poor to medicines and committing to international agreements. *Developing World Bioethics* 3: 27-48.
Concar, D. 2003. Trials and errors: With drug companies now testing new medicines in all parts of the globe, are ethical standards being compromised? *New Scientist* 178(2400): 28.
Daar, A.S., A. Bhatt, E. Court, and P.A. Singer. 2004. Stem cell research and transplantation: Science leading ethics. *Transplant Proceedings* 36(8): 2504-6.
Dhai, A., J. Moodley, D.J. McQuoid, and C. Rodeck. 2004. Ethical and legal controversies in cloning for biomedical research: A South African perspective. *South African Medical Journal* 94: 906-9.
Edejer, T.T. 2000. Disseminating health information in developing countries: The role of the Internet. *British Medical Journal* 321: 797-800.
Elliott, C. 2004. Six problems with pharma-funded bioethics. *Studies in History and Philosophy of Biological and Biomedical Sciences* 35: 125-9.
Farmer, P., and N. Gastineau. 2004. Rethinking medical ethics: A view from below. *Developing World Bioethics* 4: 17-41.
Global Forum for Health Research. 2002. *10-90 Report on Health Research 2001-2002*. Available at: <http://www.globalforumhealth.org/FilesUpld/36.pdf> (Last Accessed: 28 April 2004).
Hubbard, T., and J. Love. 2004. A new trade framework for global healthcare R&D. *Public Library of Science Biology* 2(2): 147-50.
Internet World Statistics. 2005. Internet Usage in Europe. Available at: <http://www.Internetworldstats.com/stats4.htm> (Last Accessed: 5 January 2005).
Koehn, D. 1994. *The ground of professional ethics*. London: Routledge.
Levine, R.J. 1999. The need to revise the Declaration of Helsinki. *New England Journal of Medicine* 341: 531-4.
Love, J. 2003. A new framework for healthcare R&D. Available at: <http://www.earthinstitute.columbia.edu/cgsd/documents/love_000.pdf> (Last accessed: 30 April 2004).
Lurie, P., and S.M. Wolfe. 1997. Unethical trials of interventions to reduce perinatal transmission of HIV in developing countries. *New England Journal of Medicine* 337: 853-6.
Médicins Sans Frontières, Drugs for Neglected Diseases Working Group. 2001. *Fatal imbalance: The crisis in research and development for drugs in neglected diseases*. Geneva: Médecins Sans Frontières.
Physicians for Human Rights. 2002. *Dual loyalties and human rights*. Boston: Physicians for Human Rights.
Piper, T. 2004. Commentary. *Journal of Medical Ethics* 30: 475-7.
Pollard, A.J., and J. Savulescu. 2004. Eligibility of overseas visitors and people of uncertain residential status for NHS treatment. *British Medical Journal* 329: 346-9.
Resnik, D.B. 2001. Developing drugs for the developing world: An economic, legal, moral and political dilemma. *Developing World Bioethics* 1: 12-32.
Scheper-Hughes, N. 2000. The global traffic in human organs. *Current Anthropology* 41(2): 191-211.
Schüklenk, U. 2000. Protecting the vulnerable: Testing times for clinical research ethics. *Social Science and Medicine* 51: 969-77.
———. 2004. The standard of care debate: Against the myth of an "international consensus opinion." *Journal of Medical Ethics* 30: 194-7.
Schüklenk, U., and R.E. Ashcroft. 2000. International research ethics. *Bioethics* 14: 158-72.
———. 2002. Affordable access to essential medication in developing countries: Conflicts between ethical and economic imperatives. *Journal of Medicine and Philosophy* 27: 179-95.
Schull, M. 2000. Effect of drug patents in developing countries. *British Medical Journal* 321: 833.

Sterckx, S. 2004. Patents and access to drugs in developing countries: An ethical analysis. *Developing World Bioethics* 4(1): 58-75.
Trouiller, P., P. Olliaro, E. Torrele, J. Orinski, R. Laing, and N. Ford. 2002. Drug development for neglected diseases: A deficient market and a public-health policy failure. *The Lancet* (22 June 2002).

Available at: <http://www.doctorswithoutborders.org/publications/other/miltefosine_06-2002.shtml>
(Last Accessed: 13 March 2004).
World Health Organization, Commission for Macroeconomics and Health. 2001. *Macroeconomics and
health: Investing in health for economic development*. Available at: <http://www.un.org/esa/
coordination/ecosoc/docs/RT.K.MacroeconomicsHealth.pdf> (Last accessed: 30 March 2005).

CHAPTER THREE

GEORGE F. TOMOSSY & JOLYON FORD

GLOBALIZATION AND CLINICAL TRIALS

Compensating subjects in developing countries

Globalization has contributed to fundamental changes within the biomedical research endeavor. Indeed, when considered in concert with the effects of commercialization, the transformation might be described as dramatic. With the ongoing consolidation of the pharmaceutical research industry into massive multinational corporations, the scope of the business of developing medical products has grown to unprecedented levels, spanning across national boundaries and affecting health care consumers throughout the world. The influence of the global pharmaceutical industry is both reflected in and strengthened by international treaties protecting drug patents and initiatives to harmonize regulatory requirements.[1] At the national level, industry has been shown to have a significant influence on setting research agendas, public health policy and the education and practices of health care professionals.[2] And as this chapter will argue, globalization of the pharmaceutical industry can also be felt on an individual level.

Specifically, we will address the problem that arises when a human subject from a developing country is injured or dies in a clinical trial conducted by a first-world sponsor. We submit that this scenario, which operates at the interface of developing world ethics, human rights advocacy and international law, is instrumental in examining how legal barriers may interfere with the goal of promoting global social justice.

Protecting human subjects in international clinical trials requires a range of solutions, including capacity building within developing countries to establish or improve regulatory oversight mechanisms, professional education to foster ethical conduct by investigators and political pressure to promote socially responsible corporate practices. As this chapter will contend, it is also vitally important that legal systems in first-world jurisdictions protect subjects from developing countries and permit access to justice in first-world courts to enable them to obtain compensation.

27

B. Bennett & G.F. Tomossy (eds.), Globalization and Health: Challenges for Health Law and Bioethics, 27-45.
© 2006 Springer. Printed in the Netherlands.

THE NEED FOR HUMAN SUBJECTS: ENTER DEVELOPING COUNTRIES

Human subjects are, in a sense, an indispensable commodity for achieving commercial objectives in biomedical innovation. The continued profitability of pharmaceutical multinationals depends on maintaining a pipeline of new – or, in the case of "me-too" drugs,[3] somewhat-new – drugs and medical devices that can be brought to market. The clinical trial, with both animals and humans, remains an integral part of product development. Successful clinical trials involving human subjects proving safety and efficacy are necessary in order to obtain regulatory approval before a product can be marketed. The profitability of pharmaceutical multinationals ultimately depends, therefore, on the availability of human subjects.

Fueling the development pipeline has led to a substantial increase in the volume and diversity of clinical trials over recent decades (NBAC 2001b, 4), with a corresponding increase in the demand for human subjects. Unsurprisingly, concomitant with the proliferation of international research, these trends are also evident in developing countries (OIG 2001, 6).

A number of reasons have been cited for this latter development, including: access to a larger pool of human subjects, clearer result yield and the need to accelerate approval for new drug marketing (id., 8-9). Host governments and investigators in developing countries also appear eager to attract research (NBAC 2001a, 1). However, it has also been suggested that market forces drive trial sponsors to shop around for the least expensive, least onerous regulatory environment with the lowest liability exposure, including for the purpose of avoiding potential litigation on the occasion of an adverse event (Lurie and Wolfe 1999).

At first blush, one might consider that the selection of a developing country in the course of "jurisdiction shopping" in response to market forces is not inherently problematic. Adopting the most cost-effective course of action is, after all, good business. Even preferences for a less onerous regulatory environment (which can lead to bringing a product to market faster) and for jurisdictions with lower risk of litigation can translate into cost savings. Nevertheless, furthering commercial objectives at the expense of human rights, individual well-being and moral values should engender serious concern.

An ethical, regulatory or legal concern?

The conduct of clinical trials involving human subjects in developing countries by first-world sponsors presents a host of ethical, regulatory and legal dilemmas. As with human subject protection generally, the issues raised are interrelated. Engaging with the main legal focus of this chapter on access to justice by developing world plaintiffs, therefore, requires a preliminary discussion of these related issues.

An ethical concern

Absent an appropriate allocation of risks and benefits, first-world sponsored research in developing countries would seem intrinsically unethical.

The primary ethical concern relates to the problem of exploitation, which can be defined as occurring when wealthy or powerful agents "take advantage of the poverty, powerlessness, or dependency of others to serve their own ends, without a sufficient benefit for the less advantaged individuals or group" (NBAC 2001a, 10; see also Macklin 2004, 99-130). With regard to this issue, the recently revised *International Ethical Guidelines for Biomedical Research Involving Human Subjects* produced by the Council for International Organizations of Medical Sciences (CIOMS) pronounces squarely against the commercially motivated practice of selecting developing countries for first-world sponsored research mentioned earlier:

> Sponsors of research or investigators cannot, in general, be held accountable for unjust conditions where the research is conducted, but they must refrain from practices that are likely to worsen unjust conditions or contribute to new inequities. Neither should they take advantage of the relative inability of low-resource countries or vulnerable populations to protect their own interests, by conducting research inexpensively and avoiding complex regulatory systems of industrialized countries in order to develop products for the lucrative markets of those countries.   (CIOMS 2002, 11)

The guiding principle here is distributive justice, which requires the equitable distribution of risks and benefits arising from research. Research should have direct relevance to the local communities from which subjects are drawn, and those communities should enjoy the benefits derived from their collective contribution (see generally, Macklin 2004, 68-98). Unfortunately, it has been frequently observed that only a small fraction of pharmaceutical industry expenditure on research and development has resulted in approved drugs for tropical diseases, with the balance of industry funding having been dedicated to first-world illnesses (Shah 2003, 30).[4]

Even without direct benefits accruing to the country from which subjects are drawn, benefits can be gleaned from the establishment of a clinical trial infrastructure, including the training of local investigators. Presumably, this would translate into much-needed research on diseases specific to developing countries at a local level. Such potential economic, educational and health gains should not be discounted. However, as the concern with the impact on individual subjects remains, considerations extrinsic to subjects' interests should not obviate the need for effective regulatory and legal frameworks that are geared to protect the rights of subjects, including with regard to securing their right to compensation.

A regulatory concern

Ethical concerns about distributive justice and exploitation lead to important regulatory considerations. It has been observed that developing countries often have limited capacity to evaluate and monitor clinical trials through lack of adequate regulatory infrastructure, or are faced with a substantial conflict of interest, given their strong desire to attract foreign research (NBAC 2001a, 82-3). The key question, then, is: who should be responsible for the protection of human subjects in developing countries?

An obvious solution to this problem is for developing countries to address this concern themselves. However, it would be overly harsh to place the burden of establishing systems to regulate clinical trials squarely upon the shoulders of developing countries, considering that there are far more pressing problems facing such countries and limited resources with which to respond to these problems. Nevertheless, a substantive system of research governance entrenched at the national level would indeed be the ideal solution. To that end, the Nuffield Council on Bioethics (2002, para 8.16) made the following recommendation:

> We recommend that all developing countries should have in place a properly constituted and functioning system for the independent ethical review of research. This will include the establishment of effective research ethics committees. Developing countries may determine that the most appropriate means of reviewing externally sponsored research is via an independent national research ethics committee. In such circumstances the establishment, funding and proper operation of independent national research ethics committees should be the responsibility of national governments. No research should be conducted without review at the national or local level.

Guidance for developing countries for the establishment of ethics committees has also been offered by the World Health Organization (WHO 2000; 2002), with further sources on convergent ethical principles available in the form of international guidelines, such as the World Medical Association's *Declaration of Helsinki*[5] and CIOMS guidelines (2002).

Conversely, it would also be inappropriate to demand that the country of a first-world sponsor become the singular guardian of developing world subjects. Such an approach runs afoul of concerns about paternalism or, as will be discussed below, "ethical imperialism." More practically, there may be problems with the extraterritorial application of sponsor-country regulations (Dubois 2003). This latter concern would be even greater where reliance is placed on "soft" controls, such as professional ethical guidelines, rather than official regulations. In this respect, the proposal (made in the US context) that a system of accreditation of research institutions would be "the only workable means of ensuring that American medicine does not provide benefit at the expense of the inhabitants of developing nations" (Kelleher 2004-5, 106) ought to be held in reserve for the time being. The current initiative for accreditation of research ethics committees in the United States is in its early stages, and despite having strong proponents within the US research establishment (IOM 2001), its effectiveness for protecting human subjects is yet to be assessed at the local level, let alone for participants in foreign trials.[6]

The Nuffield Council did not, of course, simply recommend that research ethics oversight should be solely the responsibility of host jurisdictions in the developing world; it also made recommendations relating to the responsibilities of first-world sponsor jurisdictions. A balanced approach, advocated by both the Nuffield Council (2002, paras 8.22-8.29) and National Bioethics Advisory Commission in the United States (NBAC 2001, 82-91), would require oversight by both host and sponsoring jurisdictions. Both bodies also recommended that first-world sponsors should contribute to capacity building in the host jurisdiction. Capacity building, which would "develop and implement strategies that assist in building local capacity for designing, reviewing, and conducting clinical trials in developing countries" (id., 91), would have the added benefit of decreasing the reliance by host jurisdictions on first-world oversight mechanisms and facilitate the pursuit of research priorities for the developing world.

Admittedly, first-world support for dual oversight is not wholly altruistic, as it would also serve to ensure compliance with standards of good clinical practice and thereby help to alleviate concerns by drug regulatory agencies about the quality of data obtained from foreign trials. Even the recommendation that first-world sponsors and researchers should participate in capacity building may also be rooted in practical considerations, as it would likewise serve to enhance the quality and reliability of foreign trial data. Such motives, however, should not be seen to detract from the benefits that dual review and capacity building might provide.

Based on a growing body of scholarship on international research ethics and thorough inquiries such as have been conducted by the Nuffield Council, National Bioethics Advisory Commission and other bodies,[7] one can safely hold out as predominant the view that first-world jurisdictions have clear ethical and regulatory obligations with respect to clinical trials in developing countries. Nonetheless, even where these obligations are satisfied, further considerations will remain of a legal nature. Owing to the level of uncertainty inherent in biomedical research, adverse events can always eventuate, despite even the most stringent system of regulatory oversight. Subjects can be injured or die in the course of a clinical trial. Thus, while regulatory measures that uphold ethical principles of distributive justice will undoubtedly contribute to the aim of protecting the rights of human subjects, account needs to be taken of issues related to compensation for subjects and the corresponding legal liability of sponsors.

A legal concern

It is undisputed that subjects are entitled to compensation for injuries sustained in the course of their participation in a clinical trial or biomedical research protocol. Research sponsors, institutions, investigators, ethics committees and even committee members may all be liable. Recent litigation in the United States has demonstrated that researchers will indeed be held to account for injuries to subjects.[8] Institutions have thus rightly been encouraged to guard against such liability by ensuring that external sponsors have "appropriate and sufficient insurance to meet [their] legal liability to research subjects for harm arising out of the particular research" (Legal Liability Working Group 1994; see also NHMRC 1999, 36). The investigator's obligation to provide compensation for injured subjects is well stated in the CIOMS guidelines (2002, Guideline 19):

> Investigators should ensure that research subjects who suffer injury as a result of their participation are entitled to free medical treatment for such injury and to such financial or other assistance as would compensate them equitably for any resultant impairment, disability or handicap. In the case of death as a result of their participation, their

dependants are entitled to compensation. Subjects must not be asked to waive the right
to compensation.

The right to compensation and corresponding legal liability would also extend to
subjects in developing countries. As the European Group on Ethics in Science and
New Technologies of the Council of Europe advocated in relation to clinical trials in
developing countries, "The standards of insurance, liability and indemnity insurance
for the participants in a clinical trial and their families must provide the same kind of
protection wherever a trial takes place" (EGE 2003, 15).

A major problem, however, from the perspective of developing world subjects, is
that clear lines of international accountability and transnational mechanisms for
enforcing a claim for compensation are generally lacking. As the following section
will show, procedural and substantive obstacles in international law can operate to
frustrate efforts to bring an action in first-world courts, effectively insulating first-
world researchers from claims arising from their activities. The developing world
plaintiff thus faces the difficult challenge of access to justice.

OBTAINING COMPENSATION: EXIT FIRST-WORLD SPONSOR

The quest for compensation by an injured subject from a developing country
intersects with the areas of tort law and international human rights law (Fidler 2001;
Todres 2000). The balance of this chapter is concerned with the latter, where the
developing world plaintiff must overcome both procedural and substantive hurdles:
  (i)   locating a mechanism to obtain a remedy under international law;
  (ii)   establishing the existence of subjects' rights under international law, which
         must occur prior to proving their breach; and
  (iii) overcoming barriers arising from the *forum non conveniens* doctrine.

The Trovan Trial litigation, which involved the families of child subjects in
Nigeria and Pfizer, a multinational pharmaceutical company based in New York,
provides an excellent case study through which to illuminate the problems caused by
the systematic operation of jurisdictional rules on transnational access to justice.

The Trovan Trial litigation

In December 2000, the *Washington Post* ran an investigative series entitled, "The
Body Hunters." It raised concerns about foreign-sponsored clinical trials conducted
in Africa, Latin America, China and Eastern Europe. The first article in this series
featured the Trovan Trial that forms the basis of this case study. The article sparked
outrage in both Nigeria and the United States, with official inquiries and litigation
having been instituted in both countries (Shah 2003, 34).

The events leading up to this litigation can be summarized as follows.[9]

In early 1996, serious epidemics of bacterial meningitis and cholera broke out in
northern Nigeria. Pfizer, a leading transnational pharmaceutical company based in
the United States, took this opportunity to conduct a clinical trial to test the efficacy
of its new broad-spectrum antibiotic, Trovan,[10] as an oral treatment for meningitis in

children. A clinical trial was quickly organized and conducted with the cooperation of Nigerian authorities. The US Food and Drug Administration expedited approval for export of the experimental drug on the basis of an apparent invitation to Pfizer from the Nigerian Government. Trovan was tested against the approved treatment that used a competitor's product.[11] It was alleged by the plaintiffs that a number of child subjects died or sustained permanent disabilities, such as paralysis, deafness or blindness, as a result of participation in the trial.

In March and August 2001, plaintiffs representing the affected children or their families brought actions against Pfizer in Kano and New York respectively. In the US action, plaintiffs relied on a federal statute, the *Alien Tort Claims Act*.[12] They pleaded that Pfizer had violated the customary international law prohibition of non-consensual human experimentation, argued to be codified in Article 7 of the *International Covenant on Civil and Political Rights*: "No one shall be subjected to torture or to cruel, inhuman or degrading treatment or punishment. In particular, no one shall be subjected without his free consent to medical or scientific experimentation." The plaintiffs alleged that Pfizer failed to explain to the parents of child subjects that the treatment was experimental, that Trovan had the potential to cause serious side effects, that participation in the study could be refused, and that conventional treatments were available at the same site free of charge from other organizations.

Pfizer steadfastly maintained a contrary position:

> The study was conducted in accordance with standard international legislation on clinical trials, and the verbal consent of family members of all treated patients... Working in countries where literacy is low, and in the middle of a life-threatening epidemic, it is not always possible to gain consent in writing, but the fact that the treatment was experimental was explained to family of every patient in two languages – English and Hausa – before the treatment was administered.[13]

The plaintiffs contended that an action in Nigeria would be bound to be decided adversely to the plaintiffs, as it alleged Nigerian government complicity in the Trovan trial. They argued that the Nigerian court system was subject to improper influence from the executive and so was too corrupt to be considered an adequate forum. The plaintiffs maintained that a US court should hear the case because Pfizer is based there. In September 2002, the US District Court (Southern District of New York) decided that it had jurisdiction to hear the case, but declined in its discretion to do so. The Court required the plaintiffs to attempt to sue Pfizer in Nigeria, which it saw as the natural or proper forum. Accordingly, with Pfizer's undertakings to facilitate an action in Nigeria, the motion to dismiss the action was granted on the basis of *forum non conveniens*.[14]

The parallel Nigerian proceedings[15] had been withdrawn or dismissed in August 2002. It was reported in the *British Medical Journal* in early 2003 that the plaintiff families had lost confidence in the Nigerian judicial system, where the case had been stalled since March 2001. They had withdrawn their action after "it became clear that they were not likely to get justice in Nigeria after their case was adjourned more than 14 times." Pfizer blamed the plaintiffs for any delay.[16]

In October 2003, the US Court of Appeals for the Second Circuit vacated the 2002 New York decision on the motion to dismiss and remanded the matter to the District Court to determine whether it was corruption or delay of the sort alleged by the plaintiffs that precipitated the dismissal or withdrawal of the Nigerian action, and to evaluate any impact on the District Court's analysis of the proper forum.[17]

Finally, in August 2005, the US District Court dismissed the plaintiffs' claim for lack of subject matter jurisdiction. The Court also held that the suit was subject to dismissal on *forum non conveniens* grounds because the plaintiffs had failed to establish either corruption or inadequate procedural safeguards in Nigeria.[18]

This litigation thus illustrates the significant obstacles faced by plaintiffs from a developing country to convince a US court to hear their case.

International law: Lack of forum

The first challenge faced by a developing world plaintiff is to locate a suitable forum in which non-state actors, such as corporations, can be held legally accountable for their activities. Corporations are not yet generally seen as having legal personality in international law,[19] which focuses on systematic abuses of human rights and on the interests and duties of states. A breach of an international norm by a non-state actor will likely be unenforceable under an international human rights forum (Dubois 2003, 203-4). For example, the Trovan plaintiffs would have little chance of success in bringing their claim through mechanisms available under the *International Covenant on Civil and Political Rights* (ICCPR). After showing to have exhausted all available domestic remedies, an individual could conceivably bring a complaint to the Human Rights Committee under the First Optional Protocol to the ICCPR.[20] However, as only states hold duties under the ICCPR, it would have to be shown that a corporation acted under state "direction, instigation or control" for that state to be held responsible (ILC 2001, 80; Crawford 2002).

It has been argued that a state could be indirectly responsible for failing to provide a remedy within its jurisdiction for wrongs by private actors committed in another state (Clapham 1993), such as by providing effective access to domestic courts (Clapham 2001; Addo 1999; Sornarajah 2001). This has been considered specifically with respect to a state's failure to regulate, control or supervise new drug research as a violation of a fundamental norm, such as the right to life (Fidler 2001, 327-9). Such arguments would seek to make obligations under international human rights law meaningful, particularly in response to the call for setting out legal responsibilities of multinational corporations in such a way that reflects their increased level of influence in a global society (Addo 1999; Koh 1991; Dubois 2003, 203-6). An important aspect of this call for accountability under international human rights law has been to mobilize national law, specifically tort law (Anderson 2002).

The course taken by the Trovan plaintiffs in their US action was to plead their claim under the *Alien Tort Claims Act* (ATCA), which enables a federal court to hear claims by non-US nationals for torts committed "in violation of the law of nations," including outside the US (Steinhard and D'Amato 1999; Pettyjohn 2003). This was interpreted narrowly by the District Court, which followed the decision in *Sosa v. Alvarez-Machain*[21] by the US Supreme Court. It held that the ATCA did not create new causes of action and only conferred on federal courts the power to hear "a narrow set of alien tort claims for violations of international law." Courts could recognize "new actionable rules based on evolving principles of international law." However, the District Court did not recognize the sources of international law advanced by the Nigerian plaintiffs. It refused to "judicially forge broad aspirational language into customary international law" and so create a private right of action, which following *Sosa* was "better left to legislative judgment." Consequently, the District Court held that the plaintiffs had failed to establish jurisdiction under the ATCA.

It is worth mentioning that a developing world subject might also bring what might be cast as an international human rights claim indirectly by invoking a recognized domestic private law category, such as the torts of battery or negligence (Scott 2001; Moran 2001, 668). Common law courts have always recognized their jurisdiction over torts committed in other countries where the defendant is subject to their control (Lowenfeld 1996). This possibility for plaintiffs such as in the Trovan trial is relatively unexplored (Swan 2001). The appeal of such an approach is that it would be unnecessary to argue the violation of a complex international human rights norm if the law of negligence, for example, would protect the same interests (Scott 2001). Apart from jurisdictional issues, it should not be overlooked that complex and country-specific rules concerning of choice of law – that is, which country's substantive tort law is applied in resolution of the dispute – might still arise.

Norms under international law

In order to pursue a claim based in international human rights law, as opposed to a simple private law tort claim, there is a need to ground the rights of developing world subjects in an accepted international law norm. While a subject's right to informed consent, for example, might be successfully constructed under customary international law by drawing upon a collection of instruments, ethical guidelines and national practices, none of the current international treaties or conventions unequivocally provides such a norm.

Arguments have been advanced both in favour (Orlowski 2003) and against (Meier 2002) whether the basic principles enunciated in the most oft-cited international instruments, the *Nuremberg Code* and *Declaration of Helsinki,* provide a source of norms under customary international law. However, it must be recalled that the *Nuremberg Code* was merely part of a decision reached by a US military tribunal[22] and has received limited attention to date in domestic courts, including in the United States (Annas 1992).[23] The *Declaration of Helsinki*, while broadly endorsed, nevertheless remains a code promulgated by a professional association (the World Medical Association). The District Court in the Trovan litigation came to a similar conclusion. It held that the *Nuremberg Code* had not been ratified by US courts nor had it been adopted by the international community. It also dismissed the *Declaration of Helsinki* as a statement of policy and lacked the specificity required under the *Sosa* doctrine. The CIOMS guidelines were similarly disregarded.

The principal international law instrument in this area is the Council of Europe's *Convention on Human Rights and Biomedicine.*[24] An Additional Protocol concerning Biomedical Research was opened for signature 25 January 2005.[25] The *Convention* is a landmark achievement towards the entrenchment of ethical principles in biomedicine within a multilateral framework. Consensus supporting its principles, however, is by no means absolute. Principles that would permit non-therapeutic research on vulnerable persons with surrogate consent, for example, attracted vehement resistance in Germany (de Wachter 1997). Indeed, at the time of writing, only thirty-two out of forty-six member states have signed the instrument with just nineteen ratifications. Notable absences among the latter include research-intensive nations such as the United Kingdom, France, Germany and Switzerland.[26] More fundamentally, enforcement mechanisms of the *Convention* are limited insofar as it contains no provisions establishing a forum or procedure for dealing with claims of those injured in the course of biomedical research. Rather, as stated in Article 31 of the Additional Protocol, "The person who has suffered damage as a result of participation in research shall be entitled to fair compensation according to the conditions and procedures prescribed by law." The Explanatory Report to the *Convention* notes that "on the topic of fair compensation, reference can be made to Article 50 of the *European Convention on Human Rights* which enables the Court to afford just satisfaction to an injured party" (Council of Europe 1997, para 146). However, this must be in relation to a decision or measure taken by an authority of a party to that convention. The *Convention,* however, was not considered by the District Court in the Trovan litigation.

Article 7 of the ICCPR, while relied upon by the plaintiffs in the Trovan case, was not of assistance. Article 7 has not to date formed the basis of any human rights action in the context of human experimentation (Dubois 2003, 203). Fidler's (2001) recent study examining whether placebo-based drug trials in developing countries violated international human rights law norms suggests little promise for potential claimants. While Article 7 of the ICCPR appears to be a stable fundamental norm at the heart of protecting the integrity and security of the person in international law,[27] it is nevertheless an open-textured norm, the content of which is subject to divergent interpretations (Swan 2001, 98; Fidler 2001, 337, 341), and thus lacks specificity. Plaintiffs will therefore face the challenge of demonstrating its breach by any particular drug trial. The Nigerian plaintiffs in the Trovan litigation were indeed unsuccessful, with the District Court refusing to recognize the ICCPR on the basis that it was not self-executing and so did not give rise to a private right of action.

It merits noting that the District Court also refused to recognize the *Universal Declaration of Human Rights,* stating that, like the *Nuremberg Code* and *Declaration of Helsinki,* it constituted an aspirational source of principle that could not impose obligations under the ATCA.

The matter is further confounded by the absence of definite accepted criteria relating to what constitutes a valid informed consent.[28] Brief expansion on this point is warranted. The requirement of a subject's voluntary informed consent is supposed to provide some protection against exploitation in research. It has been held out as a fundamental human right, being both an ethical and legal safeguard. A failure to obtain a valid consent may amount to a case of battery or negligence. Establishing the legal standard upon which to assess the validity of informed consent presents yet another challenge for developing world plaintiffs in pursuing their claims. But before one can address informed consent as a legal doctrine, one must wade through the quagmire that is the debate between "cultural relativists" and "universalists" concerning the validity of transplanting ethical principles derived from Western traditions to other cultures. Informed consent, in the Western tradition, is strongly rooted in the primacy of individual autonomy. It has been argued, however, that the consent requirement would be viewed differently in cultures where community-based rather than individual decision making is the norm. In such cases, does forcing a requirement for a Western conception of informed consent on cultures in the developing world amount to "ethical imperialism"? Or, should one adopt a position of "ethical pluralism," which holds that rules governing research practices might vary according to the country where it is carried out?[29] A satisfactory resolution of this debate cannot be provided here. However, it merits pointing out that a determination on the issue of the universality of ethical principles is a necessary prerequisite for any judicial consideration of the doctrine of informed consent as it relates to a claim for compensation by a developing world subject.

The forum non conveniens challenge

Regardless of whether a claim is pleaded under the US ATCA or as a private law tort action, a significant challenge to the developing world plaintiff lies in overcoming barriers thrown up by the common law doctrine of *forum non conveniens*. This doctrine allows a court the discretion to decline to hear a claim on the basis that a more appropriate forum exists elsewhere, irrespective of the fact that it could assert jurisdiction over the defendant (Fawcett 1995; Lowenfeld 1996; Bell 2002).

Firstly, the strong presumption in favour of domestic (US) plaintiffs for choice of forum does not apply to foreign plaintiffs.[30] The District Court in the New York Trovan proceedings held that it has "a duty to exercise restraint when assessing the sufficiency of other nations' courts and legal systems." Although the District Court at first instance took the view that Nigeria is a nation "experiencing difficulties in its transition from a dictatorship to a democracy," it nevertheless concluded at the end, having been required by the Court of Appeal to reconsider this issue, that the plaintiffs' claims of judicial corruption were not made out. For the Court, nothing in the plaintiffs' submissions on the inadequacy of Nigerian justice reached "beyond the most general of characterizations." Nor did the plaintiffs succeed in

demonstrating bias in favor of Pfizer, noting that Pfizer had lost an earlier case to
Nigerian citizens in the Nigerian Federal High Court in relation to Viagra.[31]

Secondly, the standard for deferring jurisdiction to an alternate forum is not high.
A defendant seeking to dismiss an action on the grounds of *forum non conveniens*
must demonstrate that an adequate alternative forum exists. As the District Court in
the Trovan action explained, an alternative forum is "adequate" under ATCA
jurisprudence if the defendants are subject to or amenable to service of process there
and the forum permits litigation of the disputed subject matter. It does not have to be
a perfect forum. A foreign forum will be considered inadequate where plaintiffs are
unlikely to obtain basic justice, but arguments that the other forum is too corrupt to
be effective, such as maintained by the Nigerian plaintiffs, have seldom succeeded.[32]
Not even lengthy delays owing to a country's lack of financial or judicial resources
were held by the District Court to prevent a *forum non conveniens* dismissal.

Thirdly, a plaintiff's choice of forum must satisfy a balance of public and private
interest factors.[33] A defendant, if successful in showing the existence of an adequate
alternative forum, will argue that, in determining which forum will be most
convenient and best serve the ends of justice, courts should give weight to the injury
having occurred outside the US. At first instance, the District Court concluded that
the public interest factors did not strongly support one forum over the other: Nigeria
had a "very strong interest" in the litigation, given the location of the trial and
residence status of the subject-plaintiffs; but there was also a strong public interest
in having a US Court decide issues of Pfizer's possible tortious conduct, as Pfizer
had developed, produced and performed preliminary testing of Trovan and designed
the Kano protocol in the US, all with the ultimate goal of marketing the drug
domestically. However, the District Court did conclude at first instance that the
balance of private interest factors "clearly weighs in favor of granting Pfizer's
motion to dismiss because the vexation that it would incur in pursuing the relevant
Nigerian discovery while litigating in this forum, is grossly disproportionate to any
convenience that the plaintiffs may experience." On the basis of certain undertakings
by Pfizer, including in relation to discovery of documents, the motion to dismiss the
New York proceedings was thus granted. In its final decision, the court held that,
regardless of its decision in relation to subject matter, it would dismiss the action
on *forum non conveniens* grounds, provided that Pfizer committed to similar
undertakings.

A developing world plaintiff may have good reason to seek compensation from
a first-world court: the first-world sponsor might have a subsidiary based in the
developing country that is not worth suing, having insufficient assets to cover its
liabilities;[34] where the sponsor lacks a local presence, it might be difficult to enforce
a local judgment abroad; and, as has been alleged in the Trovan litigation, plaintiffs
may have concerns about the capacity for their legal system to deliver a judgment
that is free from political bias or considerable delay. Unfortunately, the *forum non
conveniens* doctrine has made it very difficult for developing world plaintiffs to sue
a defendant corporation in its home forum, particularly in the United States. *Forum
non conveniens* decisions are often determinative of the overall outcome (Robertson

1994), as the cost of litigation often exhausts the limited legal and financial resources of plaintiffs who are unable to secure an early settlement (Meeran 2003). It is thus ironic that while new rules have evolved to support the global economy in order to facilitate investment and activities in developing countries, including in the area of biomedical research, a rather old legal doctrine continues to operate to restrict reciprocal access to justice (see also Rogge 2001, 317).

While the *forum non conveniens* doctrine may have rightly originated in order to prevent "forum shopping" in commercial and maritime disputes (van Lynden 1998), justifications based on convenience should not be applied equally to the context of personal injury litigation. Indeed, it would seem incongruous to prevent developing world plaintiffs from "forum shopping" in respect of their choice of jurisdiction to pursue a remedy when first-world corporate sponsors are themselves guilty of "forum shopping" when choosing a developing world venue for a clinical trial in the first place.

CONCLUSION

The debate surrounding the ethics of research on the transmission of HIV in developing countries served to heighten awareness about gross disparities in the availability and quality of health care resources between developing and affluent countries (de Zulueta 2001, 310). It is our hope this chapter will likewise help to stimulate debate not only on the issues relating to justice in international research canvassed here, but on the broader problem of access by citizens of developing countries to affordable medications, which represents perhaps the most poignant current global health concern.[35]

In examining the case of injured subjects from Nigeria and their action against Pfizer, we have argued that a number of obstacles exist for plaintiff-subjects from developing countries seeking to demonstrate that their injuries were wrongfully caused and obtain compensation. While the issues canvassed in the second part of this chapter relate to substantitive entitlements and procedural hurdles under international human rights law, they of course translate into problems that can be addressed ultimately only through ethical practices of investigators, sponsors and states (both developing and first-world) that are supported by properly conceived legal and regulatory frameworks.

Many of the issues raised in this chapter are neither new (e.g. Veatch 1971), nor unique to developing countries, as they represent problems that otherwise exist in first-world jurisdictions but are exacerbated by circumstances of limited resources, both in terms of regulatory infrastructure and expertise, prevalent in developing countries. In short, the problem of access to justice by developing world subjects is at once an ethical, regulatory and legal dilemma, as conveyed in the first part of this chapter. Resolving the procedural and substantive problems relating to access to justice for developing world subjects entails a two-tiered response: firstly, it requires locating a balance between the needs for improved accountability of multinational corporations for harmful conduct abroad; and secondly, it requires what may very amount to a paradigmatic shift in the legal, regulatory and ethical foundations buttressing the global research endeavour. As Benatar, Daar and Singer (2003, 138) aptly noted:

> Achieving human development globally requires more than economic growth. It also requires confronting the current challenging context of global health, developing a global mindset, basing a response on shared values, and adopting transformational approaches in governance, global political economy and capacity strengthening.

Correcting legal impediments to access to justice by injured developing world subjects, we submit, would be a small albeit important step towards achieving this goal.

NOTES

The chapter is based on a paper presented at the 28th International Congress on Law and Mental Health in Sydney, Australia (2003), and is an expanded and revised version of: J. Ford and G. Tomossy, "Clinical Trials in Developing Countries: The Plaintiff's Challenge" 2004 (1) Law, Social Justice and Global Development Journal (LGD), available at: <http://www2.warwick.ac.uk/fac/soc/law/elj/lgd/2004_1/ford/> (Last accessed: 7 April 2005). We are also grateful to Ross Anderson of the Faculty of Law, University of Sydney, for his helpful comments on earlier drafts of this chapter.

1. On the issue of treaties and drug patents, see Chapter 7 in this volume. On the trend towards harmonization of drug regulation and the promotion of mutual acceptance of clinical trial data across jurisdictions, see the International Conference on Harmonization of Technical Requirements for Registration of Pharmaceuticals for Human Use, which is discussed by Dominguez-Urban (1997). The role of industry bias in this process is critiqued by Abraham and Reed (2003).
2. Public Citizen, a non-profit public interest organization in the United States, has been on the record for a number of years regarding the incompatibility between the commercial interests of pharmaceutical companies and public interests. See, for example, a range of reports published by Public Citizen at <http://www.citizen.org/congress/reform/drug_industry/> (Last accessed: 20 April 2005), including in particular "The Other Drug War 2003: Drug Companies Deploy an Army of 675 Lobbyists to Protect Profits" (23 June 2003) and "Addicting Congress: Drug Companies' Campaign Cash and Lobbying Expenses" (7 June 2000). Such concerns appear to now have been vindicated in the recent report by the House of Commons Health Committee of the United Kingdom (2005).
3. For a discussion on the justifiability of "me-too" drugs, see Garattini (1997).
4. The disproportionate allocation of risk and benefit was the focus of an award-winning series of investigative reports in the *Washington Post* between 17-21 December 2000, entitled "The Body Hunters," and various follow-up articles available at: <http://www.washingtonpost.com/wp-dyn/world/issues/bodyhunters/> (Last accessed: 13 April 2005).
5. The World Medical Association *Declaration of Helsinki* was last amended in Edinburgh 2000, available at: <http://www.wma.net/e/policy/b3.htm> (Last accessed: 13 April 2005).
6. While it is conceded that arguments for accreditation do have merit, a proper critique of this issue and of the larger debate surrounding internal versus external controls of conduct in biomedical research go beyond the scope of this chapter.
7. A clear statement of the ethical considerations for first-world jurisdictions arising from globalization was clearly enunciated by the European Group on Ethics in Science and Technology (EGE 2003, para 2.5):

> In the context of globalisation of research, the optimal protection of the participant must be a priority no matter where a clinical trial is performed: thus, it should only be carried

out in countries with a less adequate healthcare environment, if very strict justification can be given. The more evident ones would be:
- The trial aims at addressing specific health conditions of the countries, for instance tropical diseases;
- The trial aims at addressing diseases existing also in industrialised countries but with a specifically high incidence in developing countries.
- The trial aims at developing treatments having a specific interest for the country (for instance, a new treatment cheaper than those already existing).
And in respect of the regulatory requirement of ethics committee review (id., para 2.8):
The scientific and ethical evaluation of the research protocol should be carried out by ethical committees from all countries involved. Host countries need to have a legal and ethical framework in order to take part in the clinical trial evaluation effectively and independently. The Group strongly supports EU initiatives to build local ethical committees in the host countries. It should be considered as a priority in terms of capacity building. When no local ethics committee exists, then the evaluation should be done by a mixed committee involving representatives from both EU Member States and host countries. It is essential that the members of this committee are independent and include persons representing patients' interests. If it is not possible to involve such an independent local representative in the evaluation, then no clinical trial should be implemented in the country.

8.   For a discussion of some of the recent US cases, including the highly publicized Gelsinger litigation, see De Ville (2002).
9.   It should be noted that this summary is based on the plaintiffs' statement of claim and does not constitute judicial findings of fact. The US District Court was bound by rules of procedure (for the purposes of a motion to dismiss) to typically accept the material facts alleged in the complaint as true and construe all reasonable inferences in the plaintiff's favor, see: *Grandon v. Merrill Lynch & Co* 147 F 3d 184, 188 (2nd Cir 1998). In the same way, our discussion can proceed from some assumption of the facts, since our purpose is to show the systemic operation of the doctrines that affect plaintiffs in these situations.
10. Trovan (trovafloxacin), an oral antibiotic also available in an intravenous formulation (alatrofloxacin or Trovan-IV), was first approved by the US Food and Drug Administration in December 1997 and first became available on the market in February 1998. A public health advisory was issued on 9 June 1999, by the Center for Drug Evaluation concerning the risks of liver toxicity associated with use of the drug and its recommended use was restricted to patients fitting specific criteria. See: <http://www.fda.gov/cder/news/trovan/default.htm> (Last accessed: 25 April 2005). The drug was suspended from the European market by the European Medicines Evaluation Agency in June 1999. See: <http://www.emea.eu.int/pdfs/human/press/pus/1804699EN.pdf> (Last accessed: 25 April 2005).
11. Plaintiffs also alleged that subjects in the control group (treated with an existing, approved treatment) were purposefully "low-dosed," in order to improve the comparative results of Trovan, and that this caused further avoidable injury and death. Pfizer claimed that it lost only 6% of patients in both the Trovan and Control groups, though it has been observed that this statistic might have been higher had Pfizer conducted follow-up visits to monitor the health of the subjects after completion of the study (Shah 2003, 33). Bacterial meningitis is fatal in one in ten cases and one in seven survivors is left with severe handicap, such as deafness or brain injury: see National Meningitis Association website at <www.nmaus.org> (Last accessed: 30 September 2003).
12. 28 USC at 1350.
13. The statement by Pfizer was made in response to a television documentary, "Dying for Drugs", that had been aired in the UK. See <http://www.pfizer.co.uk/template4.asp?pageid=196> (Last accessed: 25 April 2005).
14. *Abdullahi v. Pfizer Inc.* 2002 WL 31082956 (SDNY), 17 September.
15. In Nigeria's Kano Federal High Court: *Zango v. Pfizer International, Inc* (FHC/K/CS/204/2001).
16. See: *British Medical Journal.* 2003. News. 899: 326.
17. *Abdullahi v. Pfizer Inc* 77 Fed Appx 48; 2003 US App (2d Cir) LEXIS 20704, 8 October.

18. *Abdullahi v. Pfizer Inc* 2005 US Dist LEXIS 16126, 9 August.
19. *Kadic v. Karadzic* [1995] 70 F 3d 232 confirmed the potential ATCA liability of individuals for violations of public international law. Then, in *Doe v. Unocal Corporation* [1997] 963 F Supp 880, the court in an ATCA case did not rule out the possible liability of corporations for violations of public international law. In 2002, the Court of Appeals held that allegations of the Unocal corporation's close connection to Burmese governmental acts in violation of human rights were sufficient to give a court ATCA jurisdiction: *John Doe 1 v. Unocal Corp* (2002) WL 31063976. The US District Court for the Southern District of New York recently ruled, after an extensive discussion of authority including *Doe*, that a corporation is capable of being liable under international law – at least for gross human rights violations, and for purposes of the ATCA. See: *The Presbyterian Church of Sudan & Others v. Talisman Energy Inc & The Republic of Sudan* [2003] 244 F Supp 2d 289 (SDNY 2003).
20. See Article 5(2). It is also worth noting that Article 25 of the *European Convention on Human Rights* 1950 (ECHR) would not appear to preclude a non-EU citizen developing national country from bringing an individual petition to the relevant commission, and thence possibly to the European Court of Human Rights, where the trial sponsor is a State signatory to the ECHR.
21. *Sosa v. Alvarez-Machain* [2003] 266 F 3d 1045 (9th Cir 2001), *reheard en banc*, 331 F 3d 604 (9th Cir 2003), *certiorari granted* 1 December 2003. The Supreme Court confirmed (*Sosa v. Alvarez-Machain* USSC No 03-339, 29 June 2004) that the ATCA did not itself create a cause of action, so that an existing recognized tort cause of action needed to be alleged. The Court noted that while nothing prevented federal courts from developing the originally limited categories of torts recognized under the ATCA, ATCA recognition would only be given to allegations of the violation of "a norm of international character accepted by the civilized world and defined with a specificity comparable to the features" of the originally recognized but limited categories of torts.
22. The *Nuremberg Code* constituted part of the judgment resulting from *US v. Karl Brandt et al., Trials of War Criminals Before the Nuremberg Military Tribunal Under Control Council Law No. 10,* Vol. 2, Nuremberg, October 1946–April 1949. (Washington, DC: US Government Printing Office, 1949), 181-2.
23. A recent decision by the Maryland Court of Appeals is a notable exception. See: *Grimes v. Kennedy Krieger Institute, Inc* [2001] 366 Md 29, 782 A 2d 807.
24. *Convention for the Protection of Human Rights and Dignity of the Human Being with Regard to the Application of Biology and Medicine: Convention on Human Rights and Biomedicine*. Council of Europe Treaty Series No. 164 <http://conventions.coe.int/treaty/en/treaties/html/164.htm> (Last accessed: 4 May 2005).
25. *Additional Protocol to the Convention on Human Rights and Biomedicine, concerning Biomedical Research,* Council of Europe Treaty Series No. 195. Available at: <http://conventions.coe.int/Treaty/EN/Treaties/Html/195.htm> (Last accessed: 25 April 2005).
26. For information about the *Convention on Biomedicine*, including information about signatories and ratifications, see the Council of Europe website at: <http://www.coe.int/T/E/Legal_affairs/Legal_co-operation/Bioethics/> (Last accessed: 25 April 2005).
27. *Xuncax v. Gramajo* [1995] 886 F Supp 162 (D. Mass. 1995); *Wiwa v. Royal Dutch Petroleum Co* [2000] 226 F 3d 88 (2d Cir Ct App, 2000).
28. *Forti v. Suarez-Mason* [1987] 672 F Supp 1531 (ND Cal 1987), 1543; *Wiwa*, n27. On the difficulty perceived by courts in giving content to "amorphous" international law standards, see *Tel-Oren v. Libyan Arab Republic* [1984] 726 F 2d 774 (DC Cir 1984).
29. On these debates, see the exchange between Macklin (1998), Baker (1998a; 1998b) and Beauchamp (1998), and generally Macklin (1999a; 1999b). Ensuring that consent is voluntary is a related ethical concern as it can be compromised by offering undue inducements – of particular concern in developing countries where otherwise minor inducements, such as a basic medical exam or minor financial compensation, may seem exploitative (CIOMS 2002, principle 7).
30. The District Court in the Trovan litigation held that the plaintiffs, as foreign nationals with no significant ties to the district, would not to be afforded a presumption in favor of their choice of forum.
31. See: *Pfizer Specialties Ltd. V. Chyzolo Pharmacy Ltd.,* Suit No. FHC/L/CS/99, 13 Feb 2001.

32. See for example *Aguinda v. Texaco Inc* [2001] 142 F Supp 2d 534 (SDNY 2001) (World Bank report and US government statements on Indonesian judicial corruption did not render that other forum inadequate).
33. See: *Gulf Oil Corp. v. Gilbert* [1947] 330 US 235 (1947). Those factors include court congestion, unfairness of imposing jury duty on a community "with no relation to the litigation," the interest in having localised controversies decided at in that locality, avoiding problems in the conflict of laws and the application of foreign law. Private factors include ease of access to evidence, cost and convenience of witnesses attending trial, availability of compulsory process and other factors that might shorten the trial or make it less expensive. From those factors, the court must determine whether a trial would either create oppressiveness and vexation for the defendants out of proportion to plaintiff's convenience, or be inappropriate because of considerations affecting the court's own administrative and legal problems.
34. In the Bhopal Gas Disaster litigation, suing in the US was seen as necessary because the local Indian Union Carbide entity implicated in the negligence suit was in effect an empty shell: *Re Union Carbide Corporation* 634 F Supp 842 (1986). See also: Baxi and Paul (1986).
35. On the "10/90 gap," which deals squarely with this issue, see the discussion by Schüklenk and Bello in Chapter 2 of this volume. See also the inaugural issue of *Developing World Bioethics* (2002): "Drugs for the Developing World."

REFERENCES

Abraham, J., and T. Reed, 2003. Globalization of medicines control. In: *Regulation of the pharmaceutical industry*, eds. J. Abraham, and H. Lawton Smith, 82-107. London: Palgrave.
Addo, M., ed. 1999. *Human rights standards and the responsibility of transnational corporations.* Boston: Kluwer.
Anderson, M. 2002. Transnational corporations and environmental damage: Is tort law the answer? *Washburn Law Journal* 41(3): 399-425.
Annas, G.J. 1992. The Nuremberg Code in U.S. courts: Ethics versus expediency. In: *The Nazi doctors and the Nuremberg Code: Human rights in human experimentation*, eds. G.J. Annas, and M.A. Grodin, 201-22. New York: Oxford.
Baker, R. 1998a. A theory of international bioethics: Multiculturalism, postmodernism, and the bankruptcy of fundamentalism. *Kennedy Institute of Ethics Journal* 8(4): 201-31.
———. 1998b. Negotiating international bioethics: A response to Tom Beauchamp and Ruth Macklin. *Kennedy Institute of Ethics Journal* 8(4): 423-53.
Baxi, U., and Y. Paul. 1986. *Mass disasters and multinational liability: The Bhopal case.* Bombay: Sweet & Maxwell (Indian Law Institute).
Beauchamp, T.L. 1998. The mettle of moral fundamentalism: A reply to Robert Baker. *Kennedy Institute of Ethics Journal* 8(4): 389-401.
Bell, A. 2003. *Forum shopping and venue in transnational litigation.* Oxford: Oxford University Press.
Benatar, S.R., A.S. Daar, and P.A. Singer. 2003. Global health ethics: The rationale for mutual caring. *International Affairs* 79(1): 107-38.
CIOMS (Council for International Organizations of Medical Sciences). 2002. *International ethical guidelines for biomedical research involving human subjects.* Geneva: CIOMS. Available at: <http://www.cioms.ch/frame_guidelines_nov_2002.htm> (Last accessed: 13 April 2005).
Clapham, A. 1993. *Human rights in the private sphere.* Oxford: Oxford University Press.
———. 2001. Using the *European Convention on Human Rights* to protect the right of access to the civil courts. In: *Torture as tort: Comparative perspectives on the development of transnational human rights litigation*, ed. C. Scott, 513-35. Oxford: Hart Publishing.
Council of Europe, Directorate of Legal Affairs. 1997. *Explanatory report to the Convention for the Protection of Human Rights and Dignity of the Human Being with regard to application of biology and medicine: Convention on Human Rights and Biomedicine*, available at: <http://www.legal.coe.int/bioethics/gb/pdf/rapport.pdf> (Last accessed: 25 April 2005).

Crawford, J. 2002. *The International Law Commission's articles on state responsibility: Introduction, text and commentaries.* Cambridge: Cambridge University Press.

De Ville, K. 2002. The role of litigation in human research accountability. *Accountability in Research* 19(2): 17-43.

De Wachter, M.A. 1997. The European Convention on Bioethics. *Hastings Center Report* 27(1): 13-23.

de Zulueta, P. 2001. Randomised placebo-controlled trials and HIV-infected pregnant women in developing countries: Ethical imperialism or unethical exploitation? *Bioethics* 15(4): 289-311.

Dominguez-Urban, I. 1997. Harmonization in the regulation of pharmaceutical research and human rights: The need to think globally. *Cornell International Law Journal* 30: 245-86.

Dubois, W. 2003. New drug research, the extraterritorial application of FDA regulations, and the need for international cooperation. *Vanderbilt Journal of Transnational Law* 36: 161-207.

EGE (European Group on Ethics in Science and New Technologies). 2003. *Ethical aspects of clinical research in developing countries – Opinion of the European Group on Ethics in Science and New Technologies to the European Commission No. 17.* Available at: <http://europa.eu.int/comm/european_group_ethics/docs/avis17_en.pdf> (Last accessed: 13 April 2005).

Fawcett, J., ed. 1995. *Declining jurisdiction in private international law.* Oxford: Oxford University Press.

Fidler, D. 2001. "Geographic morality" revisited: International relations, international law, and the controversy over placebo-controlled HIV clinical trials in developing countries. *Harvard International Law Journal* 42(2): 299-354.

Garattini, S. 1997. Are me-too drugs justified? *Journal of Nephrology* 10(6): 283-94.

House of Commons Health Committee (UK). 2005. *The influence of the pharmaceutical industry. Volume 1.* Available at: <http://www.publications.parliament.uk/pa/cm200405/cmselect/cmhealth/42/4202.htm> (Last accessed: 20 April 05).

ILC (International Law Commission). 2001. Commentaries to the draft articles on *Responsibility of States for internationally wrongful acts* (extract from the Report of the International Law Commission on the work of its Fifty-third session, *Official Records of the General Assembly, Fifty-sixth session, Supplement No. 10* (A/56/10), chp.IV.E.2). Available at: <http://www.un.org/law/ilc/texts/State_responsibility/responsibilityfra.htm> (Last accessed: 25 April 2005).

IOM (Institute of Medicine, US). 2001. *Preserving trust: Accreditation and human research participation protection programs.* Washington, DC: National Academy Press.

Kelleher, F. 2004-5. The pharmaceutical industry's responsibility for protecting human subjects of clinical trials in developing nations. *Columbia Journal of Law and Social Problems* 38: 67-106.

Koh, H. 1991. Transnational public law litigation. *Yale Law Journal* 100: 2347-402.

Legal Liability Working Group (Australian Health Ethics Committee, National Health and Medical Research Council). 1994. *Report on compensation, insurance and indemnity arrangements for institutional ethics committees.* Canberra: National Health and Medical Research Council.

Lowenfeld, A. 1996. *International jurisdiction and the quest for reasonableness: Essays in private international law.* Oxford: Oxford University Press.

Lurie, P., and S.M. Wolfe. 1999. Proposed revisions to the Declaration of Helsinki: Paving the way for globalization in research. *Western Journal of Medicine* 171(1): 6.

Macklin, R. 1998. A defense of fundamental principles and human rights: A reply to Robert Baker. *Kennedy Institute of Ethics Journal* 8(4): 403-22.

———. 1999a. *Against relativism: Cultural diversity and the search for ethical universals in medicine.* New York: Oxford University Press.

———. 1999b. International research: Ethical imperialism or ethical pluralism? *Accountability in Research* 7(1): 59-83.

———. 2004. *Double standards in medical research in developing countries.* Cambridge: Cambridge University Press.

Meeran, R. 2003. Corporations, human rights and transnational litigation. Lecture, Castan Centre for Human Rights Law, Melbourne, 29 January. Available at: <http://www.law.monash.edu.au/castancentre/events/2003/meeranpaper.html> (Last accessed: 25 April 2005).

Meier, B. 2002. International protection of persons undergoing medical experimentation: Protecting the right of informed consent. *Berkeley Journal of International Law* 20: 513-54.

Moran, M. 2001. An uncivil action: The tort of torture and cosmopolitan private law. In: *Torture as tort: Comparative perspectives on the development of transnational human rights litigation,* ed. C. Scott, 661-85. Oxford: Hart Publishing.

NBAC (National Bioethics Advisory Commission, US). 2001a. *Ethical and policy issues in international research: Clinical trials in developing countries.* Bethesda, MD: NBAC.

———. 2001b. *Ethical and policy issues in research involving human participants.* Bethesda, MD: NBAC.

NHMRC (National Health and Medical Research Council). 1999. *National statement on ethical conduct in research involving humans.* Canberra: National Health and Medical Research Council.

Nuffield Council on Bioethics. 2002. *The ethics of research related to healthcare in developing countries.*

OIG (Office of Inspector General, US). 2001. *The globalization of clinical trials: A growing challenge in protecting human subjects.* Rep. No. OEI-01-00-00190.

Orlowski, V. 2003. Promising protection through internationally derived duties. *Cornell International Law Journal* 36: 381-413.

Pettyjohn, M. 2003. Bring me your tired, your poor, your egregious torts yearning to see the green: The Alien Tort Statute. *Tulsa Journal of Competition and International Law* 10: 513-52.

Roberston, D. 1994. The federal doctrine of forum non conveniens: An "object lesson in uncontrolled discretion." *Texas International Law Journal* 29: 353-80.

Rogge, M. 2001. Towards transnational corporate accountability. *Texas International Law Journal* 36: 299-318.

Scott, C. 2001. Translating torture into transnational tort. In: *Torture as tort: Comparative perspectives on the development of transnational human rights litigation,* ed. C. Scott, 45-63. Oxford: Hart Publishing.

Shah, S. 2003. Globalization of clinical research by the pharmaceutical industry. *International Journal of Health Services* 32(3): 29-36.

Sornarajah, M. 2001. State responsibility for harms by corporate nationals abroad. In: *Torture as tort: Comparative perspectives on the development of transnational human rights litigation,* ed. C. Scott, 491-512. Oxford: Hart Publishing.

Steinhardt, R., and A. D'Amato, eds. 1999. *The Alien Tort Claims Act: An analytical anthology.* New York: Transnational.

Swan, M. 2001. International human rights tort claims and the experience of United States courts. In: *Torture as tort: Comparative perspectives on the development of transnational human rights litigation,* ed. C. Scott, 65-107. Oxford: Hart Publishing.

Todres, J. 2000. Can research subjects of clinical trials in developing countries sue physician-investigators for human rights violations? *New York Law School Journal of Human Rights* 16: 737-68.

van Lynden, Baron C., ed. 1998. *Forum Shopping.* London: LLP.

Veatch, R.M. 1971. "Experimental" pregnancy: The ethical complexities of experimentation with oral contraceptives. *Hastings Center Report* 0(1): 2-3.

WHO (World Health Organization). 2000. *Operational guidelines for ethics committees that review biomedical research.* Geneva: WHO.

———. 2002. *Surveying and evaluating ethical review practices: A complementary guideline to the "Operational guidelines for ethics committees that review biomedical research."* Geneva: WHO.

CHAPTER FOUR

DEBORAH ZION

ETHICS, DISEASE AND OBLIGATION

> My suggestion is that privileged people need new ways of thinking about the value of
> life – not merely our own but all lives – and about the goals and purposes of medicine
> and health. Our concern needs to extend beyond the nimble crafting of paragraphs in
> complex agreements, to real concern for people!...[We need to] ask what the private
> sector (that in many ways thrives on public investments in education and basic
> innovation – as well as tax on perks) can do to justify its extraordinary level of
> expectation for the rich while the poor are marginalized.   (Benatar 2003, 71)

Twenty years ago a strange disease started killing young men in Western countries.
This disease brought about profound political and social change in the way in which
clinical research was conducted, and drew attention to the relationship between
political strength and survival. It forced many in Western countries like Australia to
finally confront issues relating to sexual difference and drug use, and how disease
was related to lack of empowerment (Zion 1995; Maenad 2003; Kippax et al. 1993).

HIV, however, is now a condition that is raging unchecked in areas where
human rights are neglected and poverty and corruption flourish. In some parts of
Africa 40 percent of persons are HIV positive. The disease brings with it copious
social and political problems, not least because of the crisis concerning orphans as
entire extended families have been wiped out (Bhargava and Bigombe 2003).

While it is clear that the drug cocktails that keep many HIV positive people alive
in Australia and other developed nations are not curative and have a range of side
effects, they do, for many people extend life for many years, long enough to bring
up one's children. The patented forms of these drugs are too expensive to be even
considered by many governments or individuals, and generic copies have, until very
recently, been unavailable because of the patent system (Reich 2000; Schüklenk and
Ashcroft 2001).

HIV/AIDS has forced people in the developed world to consider the problem of
health in the third world, not least because of the changes in clinical research and
access to drugs in countries like Australia and the US that came about as a result of
political action, particularly by politically active communities like the gay
community (Woolcock and Altman 1999). I focus upon HIV/AIDS in this
discussion of pharmaceutical companies and obligation as it is this disease that has

47

B. Bennett & G.F. Tomossy (eds.), Globalization and Health: Challenges for Health Law and Bioethics,
47-57.
© 2006 Springer. Printed in the Netherlands.

brought about this series of historical events, where a discourse on third world health, and access to life saving, if not curative drugs, has begun (Schüklenk 1998). In this discussion I set out to do four things:

1. outline the recent history of the trade agreements that deal with patents on potentially life saving drugs for HIV/AIDS;
2. discuss why drug companies and their shareholders might have obligations to people in the developing world;
3. discuss the basis upon which those who create drugs for such companies, doctors and health care workers might also have such duties;
4. reflect upon the meaning of justice.

PATENTS, DRUG PRICING AND HIV

At the end of the Second World War mechanisms for an international trade organization emerged, resulting in the formation of the *General Agreement on Tariffs and Trade* (GATT) (Loff and Heywood 2002, 623). In 1994, the World Trade Organization (WTO) was established and with it the *Agreement of the Trade-Related Aspects of Intellectual Property* (TRIPS) (Loff and Heywood 2002, 623; Shalev 2004).

Article 27 of the TRIPS Agreement requires that patents are made available for all inventions in all areas of technology, thereby mandating patents for pharmaceutical inventions. Members of the World Trade Organization discussed these issues at the WTO Ministerial Conference in Doha on 9-14 November 2001, as a result of which the *Declaration on the TRIPS Agreement and Public Health* was promulgated. Dianne Nicol (2003, 2) points out that the difficult balance between patents and access to pharmaceuticals is recognised in paragraph three of the *Declaration*, which states: "We recognise that intellectual property protection is important for the development of new medicines. We also recognise the concerns about its effects on prices."

In order to combat these problems, Article 31(b) of the TRIPS Agreement provides a mechanism known as compulsory licensing. Through this process third parties can reproduce a patented product under certain circumstances, but it is required that efforts have been made "to obtain authorization from the right holder on reasonable commercial terms and conditions and that such efforts have not been successful within a reasonable period of time." This latter requirement can be waived in the case of "national emergency or other circumstances of extreme urgency or in cases of public noncommercial use."

With these provisions in place, all seems well. HIV/AIDS can clearly be seen as an emergency. What problems then emerge?

The first of these is that the US has retreated from the agreement. In a recent news article reproduced in *Developing World Bioethics*, the US-based treatment organization Health Gap suggested that the US had tried to:

> limit licensing agreements to AIDS, TB and other so–called severe epidemics, limit importing countries to those with lowest incomes, limit exporting countries to

developing countries, limit products to medicines only and perhaps diagnostic kits and prevent pooled procurement by regional groups like the African Union.

(News 2003, 2)

However, more recently still, a deal has been reached to free up compulsory licensing (Fleck 2003, 639), whereby poor countries with no domestic manufacturing capacity might be able to import cheap drugs. However, this would be a complex procedure that in all probability would not reduce drug prices enough to make them readily available to citizens of such countries.

This recent history of resistance by drug companies to provide cheap drugs and to forgo the patenting process in the light of an international catastrophe forces us to ask two questions:

- On what basis can we ground obligations that drug companies might hold to people in the developing world?
- What might these obligations look like?

THE DUTY OF BENEFICENCE

The first argument that I will put forward is grounded in the duty of beneficence, as expressed by Peter Singer. In his formulation of the duty of beneficence, Singer (1996, 28) suggests: "If it is in our power to prevent something bad from happening, without thereby sacrificing anything of comparable moral importance, we ought, morally, to do it." Singer suggests that most of us would feel we have a duty to rescue a child drowning in a pond, thereby sacrificing only a few minutes of our time. He goes on to show how even smaller sacrifices, such as the donation of a insignificant amount of income by a well-off person, could change the lives of the severely impoverished.

Singer is responding to the argument that people close to us, either physically or emotionally, have some special claim. He argues, on the contrary, that particular attachments of this kind are irrelevant when considering our obligations to others. The only important issue, according to Singer, is that our rendering assistance does not exacerbate a bad situation, or in itself bring about comparable harm. He states:

> By "without sacrificing anything of comparable moral importance" I mean without causing anything else comparably bad to happen, or doing something that is wrong in itself, or failing to promote some moral good, comparable in significance to the bad thing that we can prevent. (Singer 1996, 28)

Singer's "obligation to assist" (Singer 1979, 128) has been criticised by many on the basis that it is unrealistically overdemanding (Arthur 1996, 43-4).

> The basis of the overdemandingness objection is simply the belief that there are limits to the demands of morality's requirements that we promote the good. Though this belief is shored up by self-interest in the case of the better-off, it is nevertheless a belief that any person might reasonably have, whatever their level of well-being. (Murphy 1993, 268-9)

When we consider global issues the concept of *organized collectivity* can help us legitimately to limit the duty of beneficence and thereby resolve some of the difficulties raised by the "overdemandingness" objection. This idea is described in detail by Joel Feinberg (1992) in his discussion of state organization and turns up again in Beauchamp and Childress, in their response to the overdemanding claims used to criticize the duty of beneficence.

Beauchamp and Childress (1989, 201) set the following limits on when persons have a duty to act:

1. Y is at risk of significant loss or damage to life or health or some other major interest.
2. X's action is needed (singly or in concert with others) to prevent this loss or damage.
3. X's action (singly or in concert with others) has a high probability of preventing it.
4. X's actions would not present significant risks, costs or burdens.
5. The benefit that Y can be expected to gain outweighs any harms, costs, or burdens that X is likely to incur.

Beauchamp and Childress' conditions can be partly fulfilled through collective action, as the load placed on each individual is thereby lessened. Acting in concert with others often increases efficacy, and this, by some accounts also strengthens claims of beneficence.

The relationship between collectivity and the duty of beneficence is particularly significant when we consider how effectively drug companies could change the lives of millions of people with a small level of personal sacrifice to individuals involved, either shareholders or company executives by reducing dividends, salaries and benefits by small amounts to already wealthy people.

The issue of the primacy of shareholders is often raised in relation to pharmaceutical companies' obligation. Thomas Pogge (2002, 127-9) has referred to the predominance of shareholders interests as the "sucker exemption." According to this argument, the protection of profits at the expense of making life saving drugs more readily available and focusing on the health needs of impoverished countries is acceptable because companies that act otherwise will not survive for long in the market, and will be replaced by those most interested in financial return (Barry and Raworth 2002, 60).

This argument has some resonance with the "replaceability defence," that is, an argument that provides an exemption from moral responsibility because an action performed by one agent would most likely be performed by someone else in the agent's absence. However, Douglas Lackey shows us how the replaceability defence does not excuse us from responsibility in his discussion of the bombing of Hue:

> Obviously if I bomb the Vietnamese in Hue in 1968, my act is necessarily a condition of the bombing…The fact that someone dropped the bombs with me does not excuse me from the responsibility of the effects of the bombs, and so still less am I excused by the postulation of a moral doppel-ganger who will drop the bombs if I do not.
>
> (Lackey 1983, 337)

Lackey argues that moral responsibility is founded on causal responsibility, and that our role in causing something to happen cannot be excused by an appeal to what might have occurred had the agent not contributed to the situation in question.

Anthony Appiah (1987, 199) argues in a similar way, and further develops this discussion by suggesting that we ask two questions:

1. Did I play a causal role in the outcome?

2. Did I make a difference to which outcome occurred?

Thus if an agent's action was a link in the causal chain, then that is enough to establish causal responsibility, even if the outcome of the situation might have been the same had the agent not acted.

Causation, while an important component of moral responsibility, is not the only significant factor. A cat may break a beloved possession and thereby be causally responsible, but this is somewhat different from another person performing the same action (Fischer and Ravizza 1993, 4-5). Our reaction will also differ if a person breaks the object deliberately or accidentally.

What then is important about the relationship between causal and moral responsibility? John Martin Fischer and Mark Ravizza (1993, 8) suggest that an agent is morally as well as causally responsible if she is knowledgable of the probable consequences of her actions.

Exposing the weaknesses in the "replaceability" defence is significant because it indicates that companies have duties and responsibilities above making money for shareholders, and that they each must bear some responsibility if their failure to uphold such duties results in suffering. This idea of corporate responsibility has some resonance with the "social entity" conception of corporations. According to Dodd (1932, 1162, as cited in Cohen and Illingworth 2003, 43):

> Business – which is the economic organization of society – is private property only in a qualified sense, and society may properly demand that it be carried on in such a way as to safeguard the interests of those who deal with it either as employees or as consumers even if the proprietary rights of its owners are thereby curtailed.

Dodd's statement is significant because it alerts us to the way in which business profit is intertwined with social conditions, and this in turn emphasizes that corporations have both moral responsibilities and obligations to those affected by their practices.

EFFICACY AND OBLIGATION

So far I have discussed the way in which the duty of beneficence might apply to pharmaceutical companies. Robert Goodin's work, *Protecting the Vulnerable*, provides additional support for this argument. Goodin (1985, 163) emphasizes the importance of collective action in relation to foreign aid and world hunger. His assertion is based upon the fact that impoverished communities suffer from problems related to poor distribution and poor infrastructure. Thus Goodin contends that when considering aid to the severely impoverished, giving money is not enough – individuals must also engage in political action to organize efficacious schemes in which governments and agencies can instigate long term change by addressing the sources of inequality (id., 164). The main advantages of collective action are, therefore, efficacy and an easing of the burden on individual donors, thus answering to some degree the "overdemandingness" objection (Zion 2004, 407).

While collectivity, therefore, has an important role to play in acceptably limiting our duties to others, limiting such duties is not Goodin's only concern. He is equally preoccupied with the way in which our actions *create* vulnerability and the kinds of duties and responsibilities we create as members of social collectives.

Goodin describes the relationship between vulnerability and responsibility in his discussion of a passenger on a plane, and the duties owed to him by the pilot. We might think that a pilot has a professional duty to his passengers, or reciprocal duties based upon payment for his services. Goodin (1985, 113), however, suggests that the fact that the passenger is especially vulnerable to the pilot's actions also creates certain special responsibilities.

It is clear that persons with HIV and other life threatening diseases are highly vulnerable to the actions of drug companies. Therefore, Goodin's argument provides us with a further reason why such companies might have obligations to people in the developing world. Christian Barry and Kate Raworth (2002, 62) make a similar argument:

> This duality of roles [by pharmaceutical companies] – as both actors within the rules and framers of the rules – is particularly clear in the case of the TRIPS agreement. Pharmaceutical company executives often accompanied national delegations in the negotiations that created the agreement, and the influence of specific companies has been well documented. While each particular company or government may legitimately claim that unilateral efforts to create access to HIV/AIDS drugs will be self-defeating, taken together they cannot so easily claim to be powerless to reshape intellectual property or other rules.

AN ARGUMENT FROM JUSTICE

My account of collectivity and beneficence goes some way to addressing why pharmaceutical companies in particular might have duties towards people in need in developing countries. I hope that this short analysis based upon the claims of justice might strengthen this argument.

In his discussion of pharmaceutical companies and the developing world, David Resnik (2001, 18) suggests that such companies have obligations because they "have obligations to avoid causing harm, and to promote social welfare." What account of justice is this in fact appealing to?

One relevant account is the idea of justice equated with compensating people for ill fortune. As Richard Arneson (1996, 1) suggests:

> The concern of distributive justice is to compensate individuals for misfortune. Some people are blessed with good luck, some are cursed with bad luck, and it is the responsibility of society – all of us regarded collectively – to alter the distribution of goods and evils that arises from the jumble of lotteries that constitute human life as we know it.

This view of justice as fair distribution is based upon the work of John Rawls, and has at its heart the idea that persons have intrinsic moral worth and entitlements regardless of their actual or potential contributions. Central to Rawls' formulation of justice as fairness is the creation of a level playing field, upon which equitable relationships can be built. He states: "All social primary goods – liberty and opportunity, income and wealth, and the bases of self-respect – are to be distributed equally unless an unequal distribution of any or all of these goods is to the advantage of the least favoured" (Rawls 1971, 303).

Shue goes further, and introduces the foundational concept of reciprocity. He also describes an important relationship between institutions and the fulfilment of positive duties to assist through the story of two men, Benny and Al. Benny is relatively wealthy, whereas Al is severely impoverished, and unsupported by the infrastructure of the welfare state. Shue asks us to consider why should Benny, who shares neither culture nor any institutions with Al, have a duty to protect and fulfil Benny's rights?

Shue's answer to this question is that these two men are indeed linked through the global economy. However, these mechanisms do not assign rights and duties. According to Shue, individuals therefore have duties "for the design and creation of positive-duty-performing institutions" in order for rights to be fulfilled (Shue 1988, 703; Zion 2004, 410).

Thomas Pogge (2002, 46) has dramatically illustrated how global institutions contribute to the ongoing cycle of poverty and corruption, through the international resource privilege, whereby they support corrupt regimes that sell off national assets, or enter into loan agreements, that must be honoured by subsequent regimes, whether the initial government was elected or not. Pogge points out that not only does this cycle encourage political instability in many parts of the world; it also keeps the prices of natural resources low for people in affluent countries (id., 48).

Pogge's analysis of how exploitation contributes to the ongoing cycle of poverty and oppression can be contrasted with a more simple view of reciprocal justice that drug companies are unquestionably entitled to the benefits that patents afford because they have taken on the burden of research and development, denying the context in which products are resourced, developed, bought and sold.

AN ARGUMENT FROM INTEGRITY

My final argument speaks to the different kinds of groups that are involved in drug company research and development. Jillian Clare Cohen and Patricia Illingworth have stated "that of the 14 drugs that (a particular) pharmaceutical company had identified as the most medically important in the last 25 years, 11 were partially supported by government funds" (Cohen and Illingworth 2003, 46). More fundamentally, the work done by drug companies would be impossible without the involvement of universities, doctors and other health care workers.

Why is this significant? It is clear that drug companies are driven by profit. Doctors, however, are supposed to have a primary focus on saving life and promoting health, not just in their own practice, but in a larger way, as a mark of belonging to a profession. It seems therefore, that the indispensable work done by medical professionals should lend strength to a claim based upon integrity that the entire way in which clinical research takes place needs to be rethought. When I refer to integrity here, I am referring to it as a public virtue. My analysis is based upon the work of Cheshire Calhoun, who suggests that having integrity means "standing for something" in the public arena (Calhoun 1995, 235). It seems reasonable that medical professionals might see the business of research as being primarily about the promotion of health and well-being rather than profit.

Integrity also has a part to play when we consider the role universities have in medical research. A recent initiative at Yale has produced guidelines to establish a framework concerning patents and licences in which university research and funding has had a part to play (t'Hoen 2003, 298). Their report begins with the following recommendation: "University research is intended to advance the common public good, a primary element of which is the advancement of health" (ibid). In order to achieve this aim, the report emphasizes that the impact of global public health should guide the kind of research undertaken, and the way in which results are disseminated. It emphasizes that university intellectual property "should be implemented in a manner supportive of developing countries' rights to protect public health" (ibid).

The Yale initiative demonstrates the way in which drug companies can be influenced by those upon whose work they depend for research and development. However, this effort should not be limited to distribution of the finished product alone, but rather inform all aspects of the clinical research process.

There are many problems that beset clinical research in the developing world. A number of these are related to the way in which research subjects already lack basic economic and political rights, and are therefore highly vulnerable before the research process begins (Zion, Gillam and Loff 2000). The term "basic rights" has been defined by the political philosopher Henry Shue (1980, 13-35) to mean rights that are foundational. That is, Shue describes rights to food and health care as foundational because without them it is impossible to enjoy other rights like education. Similarly, he suggests that rights like free speech should also be considered to be basic rights.

Hyder et al. (2004) have pointed out that a significant number of studies in the developing world do not in fact go through ethical review. This, however, is not the only problem. Jim Black, a doctor who worked in Mozambique for ten years, gives us some idea about how limited health care options are for many people in the developing world, and the effect this can have on informed consent: "even bars of soap, or transistor radio batteries, are likely to ensure trial participation" (Loff and Black 2000, 294).

The issue of informed consent is not only challenged by the desperation of those taking part in clinical trials. It is also vexed by other problems pertaining to levels of education of potential subjects, and the possibility of cultural misunderstandings among groups unused to sophisticated medical practices. More fundamentally, the agenda of research fails to address what has been called the "90/10 gap," in which 90 percent of research is carried out on conditions that afflict the world's wealthiest 10 percent.

CONCLUSION

How can we begin to address these problems? When considering clinical research and the distribution of drugs that result from the research process, the following issues must be considered (Zion, Gillam and Loff 2000):

- The research should address the real health needs of the populations that are involved in its development.
- A process should be set in place and continually monitored to ensure that involvement in research is not motivated by desperation, and that the process of the research is clearly understood.
- Follow-up studies should be made mandatory.
- The drugs that result from the research should be made available to those who need them, even if this involves some sacrifice from pharmaceutical companics.

Setting up processes to analyze the effects of clinical research and to share its benefits would go some way towards ensuring that subjects were not exploited, and that research began and ended on a more equitable basis. However, in the medium and long term, I would suggest that the way drugs are produced needs to be rethought, and in its reconceptualisation there should be some acknowledgment of the relationship between what Pogge (2002, 29) calls the deficit in "civil and political human rights demanding democracy, due process and the rule of law." One way of beginning this process might be to consider a view of justice put forth by the philosopher Elizabeth Anderson. Anderson (1999, 313) believes that that the relationships in which goods are distributed are as important as the actual distribution of the goods themselves, and these relationships must be based upon "mutual consultation, reciprocation and recognition" of the other's principles. She states

> The proper negative aim of equalitarian justice is not to eliminate the impact of brute bad luck from human affairs, but to end oppression, which by definition is socially imposed. Its proper positive aim is not to ensure that everyone gets what they morally deserve, but to create a community in which people stand in relations of equality to others.   (id., 288-9)

NOTES

An earlier version of this paper was presented at the 28th International Congress on Law and Mental Health in Sydney, Australia (2003), as part of the Preconference Symposium on "Medicine and Industry: Changing Paradigms in Health Law, Policy and Ethics."

REFERENCES

Anderson, E.S. 1999. What is the Point of Equality? *Ethics* 109(2): 287-337.

Appiah, A. 1987. Racism and Moral Pollution. *The Philosophical Forum* 18(2-3): 185-202.

Arthur, A. 1996. Rights and the duty to bring aid. In: *World hunger and morality,* 2nd edition, eds. W. Aiken, and H. LaFollette, 39-50. New Jersey: Prentice Hall.

Arneson, R. 1996. Rawls, Responsibility and Distributive Justice. Available at: <http://philosophy2.ucsd.edu/~rarneson/rawlsresponsibilityanddistributivejustice.pdf> (Last accessed: 4 May 2005).

Barry, C., and K. Raworth. 2002. Access to medicines and the rhetoric of responsibility. *Ethics and International Affairs* 16(2): 57-70.

Beauchamp, T., and J.F. Childress. 1989. *Principles of biomedical ethics.* New York: Oxford University Press.

Benatar, S. 2003. Improving global health. The need to think outside the box. *Monash Bioethics Review* 22(2): 69-72.

Bhargava, A., and B. Bigombe. 2003. Public policies and the orphans of AIDS in Africa. *British Medical Journal* 326: 1387-90.

Calhoun, C. 1995. Standing for something. *The Journal of Philosophy* 92(5): 235-61.

Cohen, J.C., and P. Illingworth. 2003. The dilemma of intellectual property rights for pharmaceuticals: The tension between ensuring access of the poor to medicines and committing to international agreements. *Developing World Bioethics* 3(1): 27-48.

Dodd, M. 1932. For whom are corporate managers trustees. *Harvard Law Review* 45(7): 1145-63.

Feinberg, J. 1992. *Freedom and fulfilment: Philosophical essays.* Princeton: Princeton University Press.

Fischer, J.M., and M. Ravizza. 1993. Introduction. In: *Perspectives on moral responsibility,* eds. J.M. Fischer, and M. Ravizza, 1-44. Ithaca: Cornell University Press.

Fleck, F. 2003. Drugs could still be costly under World Trade Organization deal. *British Medical Journal* 327: 639.

Goodin, R. 1985. *Protecting the vulnerable: A reanalysis of our social responsibilities.* Chicago: Chicago University Press.

Hyder, A.A., S. Wali, A. Khan, N. Teoh, N. Kass, and L. Dawson. 2004. Ethical review of health research: a perspective from developing country researchers. *Journal of Medical Ethics* 30: 68-72.

Kippax, S., R.W. Connell, G. Dowsett, and J. Crawford. 1993. *Sustaining safe sex: Gay communities respond to AIDS.* London: The Falmer Press.

Lackey, D. 1983. Professional sins and unprofessional excuses. In: *Moral responsibility and the professions,* eds. B. Baumin, and B. Freedman, 327-43. New York: Haven Publications.

Loff, B., and J. Black. 2000. The Declaration of Helsinki and research in vulnerable populations. *Medical Journal of Australia* 172: 292-5.

Loff, B., and M. Heywood. 2002. Patents on drugs: Manufacturing scarcity or advancing health. *Journal of Law, Medicine and Ethics* 30: 608-21.

Maenad, D. 2003. *Positive.* Crows Nest: Allen and Unwin.

Murphy, L. 1993. The demands of beneficence. *Philosophy and Public Affairs* 22(4): 267-92.

News. 2003. *Developing World Bioethics* 3(1): 2.

Nicol, D. 2003. Balancing access to pharmaceuticals with patent rights. *Monash Bioethics Review* 22(2): 50-62.

Pogge, T. 2002. *World poverty and human rights.* Cambridge: Polity Press.

Rawls, J. 1971. *A theory of justice.* Oxford: Oxford University Press.

Reich, M. 2000. The global drug gap. *Science* 287: 1979-83.

Resnik, D. 2001. Developing drugs for the developing world: An economic, legal, moral, and political dilemma. *Developing World Bioethics* 1(1): 11-32.

Shalev, C. 2004. Access to essential drugs, human rights and global justice. *Monash Bioethics Review* 23(1): 56-75.

Shue, H. 1980. *Basic rights: Subsistence, affluence and US foreign policy.* Princeton: Princeton University Press.

———. 1988. Mediating Duties. *Ethics* 98(4): 687-704.

Schüklenk, U. 1998. *Access to experimental drugs in terminal illness: Ethical issues.* Binghampton: Pharmaceutical Products Press.

Schüklenk, U., and R. Ashcroft. 2001. Affordable access to essential medication in developing countries: conflicts between ethical and economic imperatives. *Journal of Medicine and Philosophy* 27(2): 179-95.

Singer, P. 1996. Famine, affluence and morality. In: *World hunger and morality*, eds. W. Aiken, and H. LaFollette, 26-38. New Jersey: Prentice Hall.

Singer, P. 1979. *Practical ethics*. Cambridge: Cambridge University Press.

t'Hoen, E.F.M. 2003. The responsibility of research universities to promote access to essential medicines. *Yale Journal of Health Policy, Law and Ethics* 3(2): 293-300.

Woolcock, G., and D. Altman. 1999. The largest street party in the world. The gay and lesbian movement in Australia. In: *The global emergence of gay and lesbian politics: National imprints of a worldwide movement*, eds. D. Barry, J. Adam, J.W. Duyvendak, and A. Krowel, 326-43. Philadelphia: Temple University Press.

Zion, D. 1995. Can communities protect autonomy? Ethical dilemmas in HIV preventative drug trials. *Cambridge Quarterly of Healthcare Ethics* 4(4): 516-23.

———. 2004. HIV/AIDS clinical research, and the claims of beneficence, justice and integrity. *Cambridge Quarterly of Healthcare Ethics* 13(4): 404-13.

Zion, D., L. Gillam, and B. Loff. 2000 The Declaration of Helsinki, CIOMS and the ethics of research on vulnerable populations. *Nature Medicine* 6(6): 615-7.

CHAPTER FIVE

DEREK MORGAN

REGULATING THE BIO-ECONOMY

A preliminary assessment of biotechnology and law

There is a view, perhaps most famously encapsulated at least in English literature in CP Snow's novel *The Two Cultures and the Scientific Revolution* (1959), where science and scientists are thought of as occupying a place apart from the rest of the community, in cities and communities but not really part of them. Their activities are hidden – almost abstract. Films and other popular culture portrayed and to some extent still continue to portray this attitude. And it is one that to a certain extent has beset doctors, chemists, engineers, astronomers, and medical researchers. And it helps to promote what I want to try to address here, a democratic deficit in biotechnology. The second focus is the role of international law as a potential regulatory instrument in biotechnology.

We stand on the threshold of what might be thought to be a new dimension in the relationship of human sciences to biotechnology. It comes almost as a defining *constitutional* moment in that relationship. How is it made up? What controls are appropriate? What, if any, role does legal regulation or ethical reflection have to play in the new relationship between science and society? These questions form the basis of the preliminary assessment that I want to suggest in this paper.

Speaking some ten years ago, at a symposium on the challenges (both ethical and legal) presented by the rapid developments in modern genetics, British philosopher Mary Warnock (1993, 67) offered this starting point:

> Technology has made all kinds of things possible that were impossible, or unimaginable in an earlier age. Ought all these things to be carried into practice? This is the most general ethical question to be asked about genetic engineering. The question may itself take two forms: in the first place, we may ask whether the benefits promised by the practice are outweighed by its possible harms. This is an ethical question posed in strictly utilitarian form...Secondly we may ask whether, even if the benefits of the practice seem to outweigh the dangers, it nevertheless so outrages our sense of justice or of rights or human decency that it should be prohibited whatever the advantages.

59

B. Bennett & G.F. Tomossy (eds.), *Globalization and Health: Challenges for Health Law and Bioethics*, 59-69.
© 2006 *Springer. Printed in the Netherlands.*

As Roger Brownsword has pertinently observed, Warnock opens with a utilitarian, consequential perspective before qualifying it in some way with a perspective that is informed by a sense of justice, rights and decency. Warnock's approach is a combination of utilitarian and human rights thinking; pragmatism prevails (very little is categorically off limits), utilitarian benefits predominate, and attention to individual autonomy and informed consent regulate progress (Brownsword 2004, 19).

But there is a clear price to be paid for what may be thought to be the "pragmatic consensus" through which high levels of generality are achieved. The price of pragmatic consensus, indicative certainly of the current British approach to these questions, especially in the balance between individual rights and public policy, is that "again and again, as new medical developments emerge, we debate the same issues in different guises" (Brazier 1999, 167).

In place of this, I suggest, we need an operationalized notion of "lengthened foresight" and the notion of prudence that Hans Jonas (1984, 2) has bequeathed us. Attempts to articulate Jonas's "lengthened foresight" will suggest at least three familiar components:

1.   The precautionary principle.
2.   Scientific skepticism which may entail a more preventive approach by government to problems and possibilities in biotechnology, shifting the balance as Perri 6 (1998) has put it, away from curing problems to preventing them.
3.   A bi-directional notion of scientific citizenship; embracing continuing public acceptability, following transparent argument and subsequent accountability.

I want later to conclude this short review by suggesting the important addition of a fourth component.

Brownsword proceeds to identify an important limitation with Warnock's approach. *Most* can live with this synthesized regulatory perspective; but it does not of course work for the most vocal opponents of human genetic interventions, namely those who identify with what he has called the new "dignitarian alliance" – "dignitarian" because its fundamental commitment is to the principle that human dignity should not be compromised; "alliance" because there is more than one pathway to this ethic, Kantian and communitarian as well as religious. The impetus of this dignitarian alliance can be seen notably in international accords such as the Council of Europe *Convention on Human Rights and Biomedicine*, UNESCO's *Universal Declaration on the Human Genome and Human Rights*, the EC Directive on the *Legal Protection of Biotechnological Inventions*, and most recently UNESCO's *Draft Declaration on Universal Norms in Bioethics*, which is the subject of a present drafting initiative by a committee chaired by Justice Michael Kirby of the High Court of Australia. And it is with responses to biotechnology at the international level with which I want to be most concerned here.

Responding to this dignitarian opposition are those who argue that limitations to scientific advance are both undesirable and impracticable. Lee Silver, writing on human embryonic stem cell research, one of the first major biotechnological practices and possibilities to throw up what appears to be a global challenge, observes that

> the market place – not government or society will control cloning. If it is banned in one place, it will be made available somewhere else…in the end, international borders can do little to impede it.   (Silver 1999, 144)

Silver (id., 347) goes on to argue that the Roslin patent application (the patent applied for following the successful birth of "Dolly" the sheep from cell nucleus replacement technology) was purposely worded to be inclusive of human cloning so that "the inventors could use it as a legal vehicle to try to prevent this particular application from being used by anyone else." However, Silver is doubtful that the fear of patent infringement will have any effect on cloning enterprises that operate in countries that refuse to accept World Intellectual Property Organization rulings; and in any case, "the patent expires in 2017" (ibid.).

This skepticism stands in contrast to the emergence of other demands for supranational regulation of biomedicine which have come to occupy the international community in the last decade. American political scientist Francis Fukuyama (2002, 190), in his recent monograph *Our Posthuman Future: Consequences of the Biotechnology Revolution,* has observed that, one of the greatest obstacles to thinking about a regulatory regime for human biotechnology "is the widespread belief that technological advance cannot be regulated."

> Regulation seldom starts at the international level: nation states have to develop rules for their own societies before they can even begin to think about creating an international regulatory system…but there is absolutely no reason to rule out the possibility that it will emerge at this early stage in the game.   (Fukuyama 2002, 191)

What many do share however is a growing understanding of what we may call the democratic deficit in biotechnology and a growing awareness of the possibilities that international law may offer as a regulatory mechanism. It is for these reasons that I have argued that biotechnology seems to have arrived at – perhaps in part to comprise – a defining constitutional moment within, for example the UK, the EU, Australasia and in the wider international polity.

Let me illustrate this with three short quotations. First, in a recent report called *Science and Society,* a UK Parliamentary Select Committee on Science (2000, 1) observed that "society's relationship with science is in a critical phase…Public unease, mistrust and occasional outright hostility are breeding a climate of deep anxiety among scientists." This is echoed by the European Commission (2002, 4), which "believes that Europe's policy choice is not whether, but how to deal with the challenges posed by the new knowledge and its applications." And Calestous Juma of the Kennedy School of Government at Harvard, writing for the United Nations Conference on Trade and Development (UNCTAD), has observed: "Advances in science and technology have become key drivers in international relations" (United Nations 2003, 3). Ways must be found, he suggests, to provide "a forum for global consensus building in scientific issues" (ibid.).

The need for this forum arises in large part, in my view, because of the dangers of a chasm developing between science and society, mirroring Snow's *Two Cultures*. Australian judge Michael Kirby, writing over twenty years ago, cautioned of the dangers of the law failing to keep up with science. To fail to appreciate the phenomenal gravitational pull of science and technology and to chart a consequent response or even an anticipatory framework would entail societies resigning themselves he said "to being taken where the scientists and technologists' imagination leads. That path may involve nothing less than the demise of the rule of law as we know it" (Kirby 1983, 238-9).

That new institutional response is one that engages the resources of what I have called biomedical diplomacy (Lee and Morgan 2001, 305), which draws and builds upon earlier work in environmental law, and to which I shall turn shortly.

THE BIO-ECONOMY

What perhaps has until recently been missing in understanding this debate, unforeseen, and possibly unforeseeable twenty years ago when Kirby wrote, was a developing *global bio-economy*.[1] Stan Davis (building on his work with Christopher Meyer) has suggested that the next economy is gestating right now; it is the bio-economy. Biotechnology will be the next great wave after information technologies. It will begin in areas like pharmaceuticals and agriculture and, ultimately, spread throughout every economic sector, just as computers did before. The bio-economy will be characterized by pharmacogenetics, nanogenetics, globalization, and telemedicine, but alchemized in potent ways with results that we do not know, indeed that we cannot predict. In this way, the bio-economy is a paradigm reflection of the subpolitics of medicine in what Ulrich Beck (1992) has called the "risk society." I want to suggest that unless we begin to think how to regulate this emergent bio-economy, the shape and nature of which we cannot yet know and possibly not even predict, we condemn ourselves to a relationship with it that will further undermine the rule of law and disenfranchise us. Regulating the bio-economy is about identifying and elaborating a concept of scientific citizenship which will tax the resources of what I call biomedical diplomacy.

After hunting-and-gathering economies were overshadowed by agrarian economies, they in turn were displaced by industrial ones. The first began in Britain in the 1760s and the first to finish started unwinding in the USA in the early 1950s. In the 1950s and 1960s it was difficult to comprehend that computers would change every industry from manufacturing to hotels to insurance. The *industrial* economy had been in the process of being overtaken for about twenty years before the new economy was named, and we are about halfway through the information economy. According to Davis (2001), the next economy will be the bio-economy, and just as it is difficult now to see how biotechnology will alter non-biological businesses, by the third quarter of the life of next economy, somewhere in the mid-century, Davis suggests that bio-applications will be present in our non-biological lives. The Internet is the main event of the information economy's mature quarter, the last phase of it being marked by the widespread use of cheap chips and wireless technology that will let everything connect to everything else.

Of course, the structure of a primitive bio-economy was decoded in 1953, when Francis Crick and James Watson identified the double-helix structure of DNA, and completion and publication of the decoded human genome marks the end of its gestation period. We are heading into the second or growth quarter. Here, hot new industries appear, much as semiconductors and software did in the second quarter of the information economy. Thus, biotechnology will pave the way for the bio-economy era. During the next two decades, organic biotechnology will overlap with inorganic silicon infotechnology and inorganic composite materials and nanotechnologies, and many biological processes will be digitized.

The first four industries to be infused by the bio-economy era are pharmaceuticals, health care, agriculture, and food. Best known are the dozens of bio-engineered drugs already on the market. Most of these save lives by treating existing problems. One of the biggest shifts for biotechnology in the decades to come then will be the way it transforms the health care paradigm from one of treatment to one of prediction and prevention. Health care today is really sick care, which saves costs or makes money – depending on the configuration of the system – by filling hospital beds. The current, transitional managed care model operates by emptying beds. In the bio-economy, health care will work on a preventive model, making money or saving costs by helping people avoid having to enter a hospital in the first place.

Davis forecasts that beyond 2025, when we move into the mature bio-economy the effects and applications of biotechnology will spread into sectors seemingly unrelated to biology. Problems will spread as much as benefits do. Each era produces its own dark side. The industrial era was accompanied by pollution and environmental degradation. The major problem of the information age is privacy. In the bio-economy, the issue will be ethics. Cloning, bio-engineered foods, eugenics, genetic patenting, and certainty about inherited diseases are just a few of the many developments that are already illustrating this.

Represented here in the space of not much more than two or three decades is the shift from medicine to science and from science to business. The issue is not now so much about what doctors might do, but rather what scientists might enable them to do. The traditional professional structures of medicine are falling under sustained pressures of the information age. This allows the transnational corporation to promote products or services worldwide to a patient community with the means not merely to receive it but to seek it out. In this context the regulatory agenda has become filled with questions of competition law, intellectual property rights, licensing and registration. Ulrich Beck (2000) has suggested that we may be on the edge of a second modernity. While the postmodernist challenged scientific rationality,[2] in the second modernity the transnational enterprise may wrest back control, evading the grasp of any single jurisdictional regime. And, with the increasing precision of genetic targeting in pharmacogenetics and the related area of pharmacogenomics, more importance attaches to questions of social acceptance, which consequently move upstream.

The European Commission has neatly caught the temper of the choices that we are beginning to face. Life sciences and biotechnology are widely recognized to be the next wave of the knowledge-based economy. Europe is currently at a crossroads (European Commission 2002, 4). *Mutatis mutandis* the wider international community. One particular difficulty, identified by Linda Nielsen and her colleague Berit Faber (2002, 22), is that "ethical concerns are difficult to document and might be perceived as obstructing the promotion of international trade." Yet, with the development of biotechnology and the emergent bio-economy comes the opportunity, almost the responsibility, to continue the process of defining the status of the EU and other political unions as a polity *beyond their economic bases.* This arrives, hardly surprisingly, at a time when in international law generally, in international human rights law specifically, and in a drilled down fashion, in domestic legal systems more particularly, the responsibility of state and non-state actors to secure more than economic ideals is being actively debated.

However, in relation to human genetics, some legal regimes are *already* characterized as being "pro-regulation," others as being "anti-regulation." European regimes and jurisdictions such as Australia tend to be put in the former category, the United States in the latter. Not every State, of course, will be committed to a regulatory model based upon advance determination of what is to be permitted. This may be because of an instinctive political commitment to regulation primarily through the market, or because of an inability to reach a political commitment of any complexion.

Of course, wise government does not always legislate at the first opportunity, and the global nature of the biotechnological revolution makes the lack of attention to concerted international legislation perhaps less surprising than would be its presence to date (Fukuyama 2002). Indeed, there is sometimes a temptation to believe that legislative attempts to secure recognition of one particular view at the expense of others is the enforcement of moral majoritarianism. Thus legislation is sometimes asked to portray or reflect a weakened and less expansive ethical or moral conception.

It is the possibility of negotiating any supranational norms to which I now want to turn. And, in doing so, I want to address two foundational issues; they are of *process* – how could we achieve such international norms; and *rationale* – why would we want to try and how could we justify this anyway?

ON PROCESSES

Fukuyama (2002, 191) again is useful for my purposes in suggesting some of the initial processes involved, and it is worth repeating his observation that regulation seldom starts at the international level: nation states have to develop rules for their own societies before they can even begin to think about creating an international regulatory system. As the Danish Council of Ethics suggests, regulation concerning biotechnological developments must take its point of departure in national conditions (Nielsen and Faber 2002, 22). The Council argues that the acquisition of knowledge can in some instances raise ethical questions, and it is important that such questions be discussed internationally in order, if possible, to create consensus. But this does not exclude national regulation. An important issue, of course, is whether adoption by, for example, Denmark of its own legislation, means a greater possibility of influencing the supranational law, or whether it retards that possibility.

But evidently, an international consensus on the control of new biomedical technologies will not simply spring into being without a great deal of work on the part of the international community and the leading countries within it. There is no magic bullet for creating such a consensus, even if that were to be desirable. It will require the traditional tools of diplomacy: rhetoric, persuasion, negotiation, and economic and political leverage.

This is what I have suggested might be called *biomedical diplomacy*. This would have a number of tasks of which for now the most interesting are charting the intellectual history and identifying responses to and possible fora in which to respond to demands for supranational regulation of biomedicine; and secondly, identifying increasing regulatory sites and practices of biomedicine and justifying them.

But there are in this four cautions, and they represent the most urgent of the preparatory intellectual tasks to which I think we would want to attend.

First, and most important from the recent work of Scott Barrett in his game theoretic approach to environmental treaty making in his book *Environment and Statecraft* (2003), is the observation that without highly developed methodological and epistemological planning, international treaty making in environmental law does not – with the notable exception of the *Montreal Protocol on Ozone Depletion* – actually achieve the effects often hoped for. Thus, a detailed set of limited, specific objectives with clearly identifiable gains for all states complying with treaty obligations would be a prerequisite for international agreement in biotechnology. And we have already seen in the UN attempts to begin negotiations on a global prohibition on human cloning how difficult that is going to be.

Secondly, biotechnology does not step unaccompanied and unheralded onto the regulatory stage. From the 1960s onwards, governments in at least all Western countries have become caught between the public's resistance to paying more tax, leading to the abandonment of an interventionist role and the "hollowing out" of the state, and their public's continuing demand for high quality health, welfare infrastructures and social order, characteristics even of Robert Nozick's (1974) minimalist night watchman state. The public demand for engagement with biotechnology, both as consumer and as critic, and its possible health consequences, needs to be heeded and addressed.

A third dynamic of which any account of modern biotechnology policy must be aware is what might be called the "shelling" of the state; the removal or replacement of some or many of its powers or scope for action by the changing economic environment in which states act. The fluid organization and structure of many modern companies; the process of internalizing markets,[3] the recognition of multinational firms as efficiency seeking firms,[4] combine to draw more countries into the global economy *via* the activities of multinational corporations. Writers place different degrees of emphasis on the processes. Some talk about the end of the state (Ochame 1990; 1996), whilst others talk more in terms of the changing role of the state in an era of globalization, where states find that their role has become increasingly complex in a more highly interdependent world and take on the role of attempting to attract international capital rather than of regulating it (e.g. Strange 1988; Cerny 1990, 237; Roscrance 1996).

Finally, these interconnected developments provoke many to believe it is now necessary to bring multinational corporations under the authority of international human rights frameworks; the changing international legal system and the effect of economic globalization in undermining the significance of the state as controlling actor need, on these accounts, to be rethought. According to David Held (2001):

> Challenges to the role and primacy of the state in world politics has, however, fundamentally altered the role and function of international law, which in recent years has expanded to concern itself not only with relations between states but also, through concerns about human rights, with the individual.

These twin forces of economic globalization and reshaped government responsibilities propose very different possibilities for *individual* states and their responses to the requirement of governance in bioethics. The use of traditional legal controls and sanctions is waning and the desirability of establishing some predictive normative requirements growing. During the course of the twenty-first century, commercial and international lawyers will be asked to address whether the activities of transnational corporations that violate human rights will be subject to international and legal regulation.

But, the opportunity and the costs for states to act alone are diminishing. One simple problem suggested by Fukuyama (2002, 199) is that "if there is any region of the world that is likely to opt out of an emerging consensus on the regulation of biotechnology, it is Asia." A number of Asian countries either are not democracies or lack strong domestic constituencies opposed to certain types of biotechnology on moral grounds. The immediate problem with this observation is that, as the recent legislation on stem cell research in South Korea suggests,[5] it is overly pessimistic. But, because we may yet agree with Fukuyama's view (2002, 192) that "in the future, biotechnology may become an important fracture line in world politics," it is as a response to that that biomedical diplomacy should be concerned.

An important developmental vector is what has become known in administrative and public law literature as the concept of "smart regulation." This entails an understanding that smart regulators know that traditional command and control law (criminal law) does not always have the intended effect (Brownsword 2004), that private law remedies are of limited impact, and that public law control exercised by agency licensing or negotiation is open to the twin charges of being too soft or being too tough (ibid.). In this account, as Roger Brownsword has again observed, even smarter regulators know that they can sometimes achieve the desired regulatory effect by relying vicariously on non-governmental pressure (whether in the form of self-regulation or co-regulation by or with business or the professions, pressure exerted by consumers, the activities of pressure groups, and so on) or by relying on market mechanisms; in addition, they know that careful consideration needs to be given to selecting the optimal mix of various regulatory instruments (ibid.). As for the very smartest of regulators, they know that conduct is sometimes most effectively channeled by relying, not on norms, pressure, or financial signals, but on "code," on integrated technology or design, that is, as Lawrence Lessig (1999, 53-4) puts it, on a West Coast rather than a (traditional) East Coast approach to regulation. The aim of this new version of smart regulation is not only to understand the ways in which alternatives to law regulate, but also to understand how law might be used to make selections among these alternatives.

WHY BOTHER? HOW TO JUSTIFY?

So much for process, but what in conclusion about rationale?

Fashioning responses to these challenges of the bio-economy is the first and perhaps most central task for biomedical diplomacy. It is one to which the study of medical law, as part of a humane reflection on science, must both urgently attend and contribute. If our understanding of medicine's task is to be driven by our understanding of the human values at stake, of which of course ethics is only one – medical (or health care) law will need to pay very careful attention to the warning sounded by theologian Alastair Campbell in his Tuohy lectures of 1994, in which he articulated a vision of modern post industrial society that attempted to see health "as best understood as an aspect of human freedom" and the essence of good health care as "a liberation, a setting free" (Campbell 1995, 2).

Campbell urged attention to the fundamentals, ones that he suggested modern bioethics was in danger of being drawn away from by "the dazzle of modern medicine's alleged success." He urged that we should "try to see things from the perspective of those who suffer and for whom there is no quick technological fix." His aim insists that we focus attention on those most often overlooked in such debates, "that we listen first and foremost to the voices of those who suffer" (ibid.). The reason for this is simple, and in my view compelling:

> I do not claim that listening to the voices of the dispossessed is all that needs to be done about health and health care...but I do insist that unless we confront these issues of freedom, oppression and liberation, we will have missed the central problem of modern health care ethics. We will be focusing our attention on secondary issues while refusing to confront or even acknowledge the primary issue – that of the uses and abuses of power in health care delivery and in the very definition of health itself.   (ibid.)

Take the illustrative question of patenting and biotechnology. Here, there has indeed been much discussion of the ethics of patenting, but far less of the empirically significant question of its economics. In the ethical discussions principles of allocative justice have predominated at the expense of an equal consideration of aspects of distributive justice; there has been a concentration on *what* should be protected rather than *who* should benefit from that protection. "Marketization" of biotechnology foregrounds these issues of power and distribution at various levels and one of the severest tests for biomedical diplomacy is that of "rethinking equity in health."

Unless these distributional issues are addressed we cannot be assured that bio-ethics let alone biotechnology law will remain relevant in the bio-economy, and all talk about human values, human dignity, human rights and democratic balance will be so much empty rhetoric.

NOTES

An earlier version of this chapter was presented at the 28ᵗʰ International Congress on Law and Mental Health in Sydney, Australia (2003).

1. I have traced some early aspects of this in an essay with my colleague Bob Lee: see Lee and Morgan (2001). For an introduction to the bio-economy see Davis (2001, 179).
2. There is an excellent account and collection of this literature in Komesaroff (1995), especially in the essay by Redding (1995), and in Hyde (1997).
3. On the extent to which companies invest capital offshore and establish wholly or jointly owned subsidiary companies overseas rather than just sourcing goods via systems of international trade, see Caves (2000).
4. On spreading their production processes worldwide in order to take advantage of lower production costs or particular competitive conditions of a national market, see Dunning (1993, 79-80).
5. *Bioethics and Biosafety Act*, Act No.7150.

REFERENCES

Barrett, S. 2003. *Environment and statecraft: The strategy of environmental treaty-making.* Oxford: Oxford University Press.
Beck, U. 1992. *Risk society: Towards a new modernity.* Translated by M. Ritter. London: Sage.
————. 2000. *What is globalisation?* Cambridge: Polity Press.
Brazier, M. 1999. Regulating the reproduction business? *Medical Law Review* 7: 166-93.
Brownsword, R. 2004. Regulating human genetics: New dilemmas for a new millennium. *Medical Law Review* 12: 14-39.
Campbell, A. 1995. *Health as liberation: Medicine, theology and the quest for justice.* Cleveland: Pilgrim Press.
Caves, R.E. 2000. The multinational enterprise as an economic organization. In: *International political economy: Perspectives on global power and wealth*, 4ᵗʰ edition, eds. J.A. Friedan, and D.A. Lake, 144-55. New York: St. Martin's Press.
Cerny, P. 1990. *The changing architecture of politics: Structure, agency and the future of the state.* London: Sage.
Davis, S. 2001. *Lessons from the future: Making sense of a blurred world.* Oxford: Capstone Press.
Dunning, J.H. 1993. *Multinational enterprises and the global economy.* Essex: Addison Wesley.
European Commission. 2002. *Life sciences and biotechnology: A strategy for Europe.* Communication from the Commission to the European Parliament, the Council, the Economic and Social Committee and the Committee of the Regions. COM(2002) 27.
Fukuyama, F. 2002. *Our posthuman future: Consequences of the biotechnology revolution.* London: Profile Books.
Held, D. 2001. Violence, law and justice in a global age. In Social Science Research Council Essays on Globalization after September 11. Available at: <http://www.ssrc.org/sept11/essays/held.htm> (Last accessed: 2 May 2005).

Hyde, A. 1997. *Bodies of Law*. Princeton: Princeton University Press.
Jonas, H. 1984. *The imperative of responsibility: In search of an ethics for the technological age*. Chicago: University of Chicago Press.
Kirby, M. 1983. *Reform of the law*. Oxford: Oxford University Press.
Komesaroff, P.A., ed. 1995. *Troubled bodies: Critical perspectives on postmodernism, medical ethics and the body*. Melbourne: Melbourne University Press.
Lee, R., and D. Morgan. 2001. Regulating risk society: Stigmata cases, scientific citizenship and biomedical diplomacy. *Sydney Law Review* 23: 297-318.
Lessig, L. 1999. *Code and other laws of cyberspace*. New York: Basic Books.
Nielsen, L., and B. Faber. 2002. *Ethical principles in European regulation of biotechnology: Possibilities and pitfalls*. Copenhagen: Biotek.
Nozick, R. 1974. *Anarchy, state and utopia*. Oxford: Blackwell Books.
Ochame, K. 1990. *The borderless world: Power and strategy in the interlinked economy*. London: Profile Business.
————. 1996. *The end of the nation state*. London: Harper Collins.
Perri 6. 1998. *Holistic government*. London: DEMOS.
Redding, P. 1995. Science, medicine and illness: Rediscovering the patient as a person. In: *Troubled bodies: Critical perspectives on postmodernism, medical ethics and the body*, ed. P.A. Komesaroff, 87-112. Durham: Duke University Press.
Roscrance, R. 1996. The rise of the virtual state. *Foreign Affairs* (July/Aug) 45-61.
Silver, L. 1999. *Remaking Eden: Cloning, genetic engineering and the future of humankind*. London: Phoenix.
Snow, C.P., 1959. *The two cultures and the scientific revolution*. New York: Cambridge University Press.
Strange, S. 1988. *States and markets*. London: Pinter.
United Kingdom Parliament, Select Committee on Science and Technology. 2000. *Science & society – Third report*. London.
United Nations. 2003. *Science and technology diplomacy: Concepts and elements of a work programme*. New York.
Warnock, M. 1993. Philosophy and ethics. In: *Genetic engineering: The new challenge*, eds. C. Cookson, G. Nowak, and D. Theirbach, 67-74. Munich: European Patent Office.

CHAPTER SIX

ALAN IRWIN

THE GLOBAL CONTEXT FOR RISK GOVERNANCE

National regulatory policy in an international framework

Since the 1990s, globalization has been the focus of an enormous amount of political and social scientific discussion (e.g. Friedman 2000; Held et al. 1999). This chapter sets out to explore one important aspect of the "global" phenomenon: the relationship between, on the one hand, internationalized patterns of innovation and development and, on the other, national policy processes.

This relationship is especially significant for risk management and regulatory policy. The tendency has broadly been for individual nations to act (or at least talk) as if they are in charge of their own destinies – especially when addressing domestic audiences. Certainly, public debate and discussion over controversial areas of science and technology have generally assumed a national focus. However, the fact cannot be avoided that the specific innovations under discussion (pharmaceuticals, agrochemicals, nanotechnology) and the wider issues being raised (medical ethics, corporate social responsibility, health and the environment, the costs and benefits of new products) are international in their development and application.

Equally, global regulatory regimes have emerged for many areas of risk and health protection so as to ensure that countries do indeed act together and that barriers to trade are avoided. Indeed, for many issues affecting European nations, international regulation through the European Commission has largely replaced national regulatory activities. Going further, Millstone (2002) has observed that responsibility for setting regulatory standards covering protection of the environment and public health has steadily shifted in Europe from nation states to European Community-wide institutions to global bodies such as the Codex Alimentarius Commission and the World Trade Organization. For industry in particular, such a "Europeanized" – or even globalized – approach has obvious advantages. However, it can mean the development of very complex legal and regulatory structures which typically combine regional, national and international duties and responsibilities.

B. Bennett & G.F. Tomossy (eds.), Globalization and Health: Challenges for Health Law and Bioethics, 71-85.
© 2006 *Springer. Printed in the Netherlands.*

Rather than engaging with this discussion at an abstract and general level, this chapter will explore issues of global and national governance as they have developed within one regulatory case: that of genetically modified (GM) food in the United Kingdom and, specifically, the decision over the commercial growing of GM crops in the UK. Given the global development of GM, its implications for worldwide trade, and the growth of international mechanisms (including the EU) for regulation and control, GM illustrates nicely the challenges of governing risk in a globally interdependent process. Is it even possible for a single nation to stand alone on this issue?

Discussion here will not attempt to resolve the question of whether the UK policy process has really been swept aside by international (specifically, European) processes since such a unitary judgment would inevitably ignore the more interactive and cross-cutting factors that are at the heart of this paper. Instead, the focus will be on the role of global forces in making national policy. How do global concerns enter national policy processes? What form do these concerns take and at what points do they arise? Rather than suggesting that globalization is an either/or judgment (either global or national), I will consider the contemporary interaction of "global" and "national" factors within scientific governance.

This chapter will specifically challenge the conventional notion that globalization is simply an external force which impacts upon national processes and practices. Instead, I will explore the different ways in which "the global" takes shape within, in this case, national decision making over the direction of scientific and technological innovation. Far from being a coherent or unidirectional force, globalization should instead be seen as an interaction (or negotiation) between regional, national and international systems, and, very importantly, between different forms of socio-technical discourse.

In presenting this analysis, I will be supporting Yearley's argument that the study of globalization is also the study of certain "universalizing discourses." As Yearley (1996, 100) puts it: "These universalizing discourses have held out the prospect of resolving apparently intractable global problems by providing us with the tools to describe and analyze these problems in objective and authoritative ways." Two such discourses will be especially important in the following discussion: the discourse of economics and that of science. However, my consideration of the GM case will start with a third discourse which is increasingly being presented as a route to conflict resolution and consensus formation – that of public engagement and consultation (see also Irwin forthcoming; Hagendijk 2004).

CONSTRUCTING THE GLOBAL IN A NATIONAL SETTING: THE UK DIALOGUE ON GENETIC MODIFICATION

In the late 1990s, the UK government found itself in a difficult position over GM food. Monsanto's decision to release GM soya onto the European market in 1996 had substantially raised the profile of the whole issue and crystallized public opinion. Newspaper references to "Frankenstein foods" began to appear.

Meanwhile, EU consent already existed for the commercial cultivation of a small number of GM crops – putting national governments under pressure to allow plantings to go ahead. Under EU law (the Deliberate Release Directive of 1990), delay to product distribution could only occur if a member state had grounds for suspecting specifiable risks to health or the environment: a suspicion that required scientific evidence. As the Agriculture and Environment Biotechnology Commission (AEBC) described the situation in 2001:

> The legislation does not allow the European Commission or any Member State to turn down an application on grounds other than those specified in the Directive. Consent could not, for example, be refused on the grounds of public concern about the technology in principle. Public concerns which go beyond the criteria prescribed for the regulatory arrangements have no expression in this process…This has caused some tension between and within Member States, and has contributed to the seizing up of the EU regulatory system. (AEBC 2001, 9-10)

In 1998, and in apparent response to this requirement for scientific support for any concerns over GM licensing, the government announced the establishment of a new research program: Farm Scale Evaluations (FSEs) of genetically modified herbicide tolerant crops. At the same time, the UK government reached a voluntary agreement with industry that no GM crops would be grown commercially in the UK until the results of the FSEs had been assessed. The specific focus of the FSEs was on the comparative impact of GM and "conventional" crops on the abundance and diversity of farmland wildlife. As Grove-White (2002) has described the FSEs: "So the farm-scale trials were born – a scientific programme, but one shaped by legal constraints and political contingencies." Meanwhile, the EU regulatory system did indeed seize up as the ad hoc moratorium on approving GM food for sale became established. This was to last from 1998 until 2004 when a strain of GM sweet corn was eventually approved for human consumption (Browne 2004, 38).

In its 2001 report, the AEBC (of which Grove-White is a member) argued that, although the FSEs were expected to produce useful data that could contribute to a decision as to whether the crops involved should be commercialized, they were "not a sufficient condition for reaching those decisions" (AEBC 2001, 13). In addition, the AEBC called for a wider debate over the commercialization of GM in the UK. Crucially, this should include "informed public discussion" of the issues:

> We believe that the Government must now encourage comprehensive public discussion of the ecological and ethical – including socio-economic – issues which have arisen. Time is needed for people to overcome differences of language and explore the extent of their shared understandings, and above all there is a need to include those who have felt themselves to be excluded and hence to have no control over events. (id., 12)

In May 2002, the UK Secretary of State announced that the government was inclined to accept the AEBC's recommendation for a full debate and that public discussion would be one part of this. The planned "public dialogue" would consist of three strands:

- a public debate, managed by its own steering board;
- a study into the overall costs and benefits of GM crops, to be conducted by the Prime Minister's Strategy Unit;
- a scientific review of the issues relating to GM crops and food, conducted by a panel of independent scientists.

The plan was that all three of these streams would report publicly in 2003. Alongside the results of the FSEs, these would then assist the UK government in deciding what one key minister described as "the overall direction of its policy towards GM crops" (Beckett 2003, 4). The Government's formal response to the GM dialogue duly appeared in March 2004 (DEFRA 2004a). As the Secretary of State summarized her position at the end of the dialogue:

> There is no scientific case for a blanket approval of all the uses of GM…But equally there is no scientific case for a blanket ban on the use of genetic modification. I know of no one who argues, for instance, that the GM tool can alone solve the problems of the developing world. But it is less than honest to pretend…that GM has not the potential to contribute to some solutions.   (DEFRA 2004b, paras 26-7)

In the very framing of the GM dialogue and the FSEs we can identify the cross-cutting national-global relationship identified in the previous section. The construction of the dialogue as a British exercise suggests a high degree of autonomy and independence. However, the background to the dialogue was largely constituted by the EU regulatory framework – and specifically the requirement for appropriate technical evidence. In this situation, it was possible to present the whole dialogue as an elaborate charade: an attempt at legitimation made necessary by the UK's ambivalent relationship to the European Union. Of course, it was also possible to present the dialogue as a legitimate attempt at national influence over a global system.

In what follows, I will use the three-stranded GM dialogue as an opportunity to explore the manner in which global discussions have become embedded in national debate (for reasons of space, I will not consider further the FSEs). The GM dialogue offers an excellent opportunity to compare the social construction of the global across three different forms of "global talk." What form has "the global" taken within discussions over GM? Who speaks for the global and in what ways? This analysis will be followed by a consideration of the conceptual and practical consequences for our general treatment of national-global issues, and, more specifically, for regulatory policy.

Strand one: The public debate

Turning first of all to the public debate strand of the dialogue, the official report on this pulls together a very broad range of public submissions – including over 675 meetings, 2.9 million website hits, and over 1,200 letters or emails. The report presents issues of global governance very prominently in the context of general opposition to GM:

> GM is a very broad issue. It is not only an issue in its own right but acts as a proxy for many other current concerns which provoke strong feelings. GM makes people examine

not only its direct implications but also such issues as their own power as citizens and consumers to determine the future of their food and their environment, the influence of big business over national governments, science and research, and the media, the future of developing countries and the power of the United States and international bodies such as the World Trade Organization thought to be under its influence.

(GM Nation? 2003, 19)

More specifically, and in line with the previous discussion, anti-GM arguments as represented within this report employ the language of "the global" in several overlapping ways:

- by pointing out the lack of local or national autonomy – notably, in terms of global forces constraining the choice to adopt an organic future: "Could nations, regions and local authorities declare themselves GM-free zones or introduce their own controls on GM produce, beyond the terms of UK or EU policy or the rules of the WTO?" (id., 20);
- by opposing the power of multinational companies – and especially what is seen as a move to "secure monopoly control within the food chain, to reduce farmers to dependency…and to gain control over consumers and even governments" (id., 21);
- by arguing the case for, as the report put it, "doing right" by developing countries. Oppositional arguments suggested that GM would encourage monoculture, and lead to the loss of local knowledge and skills. The argument was also made that, whilst GM crops might have some value, other approaches would have greater impact: for example, land reform, fair trade, more sustainable farming methods; and
- by suggesting that international experience of GM was not positive. Arguments here suggested disenchantment with GM – especially among United States and Canadian farmers.

Meanwhile, pro-GM arguments, which were very much in the minority within the debate, cited a smaller number of global arguments:

- the benefits of GM to developing countries (notably, India and China); and
- the international competitive advantage to the UK of developing GM (and, more pointedly, the competitive disadvantage of not developing GM).

Finally, global arguments entered at a very broad level when the very existence of the GM debate was compared to the UK government's apparent refusal to listen to public opinion over one other major international issue: the Iraq war. As the report puts it, the independence of the GM debate was queried by those who saw very little open-mindedness in the UK government's response to public criticism of the war. In other words, "both suffered from a degree of guilt by association" (id., 23). If nothing else, this connection to the Iraq war reminds us that issues of risk and regulation cannot be separated – at least in terms of public evaluation – from other global concerns.

Taking these points together, it would appear that issues of the global did indeed arise within the public discussion – chiefly, but not exclusively, from those opposed to the commercialization of GM in the UK. However, some arguments around the global were used both to support and oppose GM so that, for example, the case of developing countries could be used in either way. What is also apparent is that public awareness of global issues, and specifically of the power and influence of multinational companies, affected the evaluation of GM technologies: their construction as "global" (often meaning in this context "North American") encouraged a negative evaluation by many members of the public. The public discussion phase of the dialogue reveals that the language of "globalization" is not always neutral but can carry heavy ideological baggage: for many public participants the very labeling of GM as a global issue led to it being viewed skeptically.

Strand two: The economic analysis

Turning to the economic evaluation of the costs and benefits of GM crops, issues of the global are also very apparent:

> The UK does not take decisions in isolation; it is embedded in an international context that includes the European Union, international rules on trade, multilateral environment agreements and links with trading partners and other overseas bodies…In a globalizing world, and with a technology as mobile as GM crops, the actions taken by the UK will have repercussions at the international level.   (Cabinet Office Strategy Unit 2003, 78)

Several examples of this global context are offered within the economic report:

- GM crop regulatory policy: The framework of international regulations is summarized by the report – including the role of the EU, the World Trade Organization (WTO), the Cartagena Protocol on Biosafety (signed by 103 countries) and the Codex Alimentarius;
- Agricultural policy and the environment: UK agriculture is presented as being strongly influenced by international developments;
- Science, innovation and competitiveness policy: Science and technology are important for UK productivity and competitiveness, and GM crops are presented as a likely test case for the UK – particularly in the context of a "competitive international environment in which other countries are also seeking to support scientific research in order to develop future prosperity" (id., 48);
- International development policy: "Considerable controversy" is noted about the role of GM crops in developing countries;
- International impacts: The report notes that UK policy towards GM will have "international repercussions" and that "[d]ifferences in approach between the EU and some other countries on GM policy are already causing trade tensions" (id., 13). The report goes so far as to suggest that "[t]his case risks not only generating fresh antipathy over GM crops, but could also have serious knock-on implications for trade relations and international law more generally" (id., 81);
- Impacts on the UK science base: It is observed that the UK government has an overall aim of expanding research and exploiting research outputs – a

deterioration in the climate for GM crops could, for example, reduce multinational research activity in the UK.

In noting these general issues, the economic report was eager to argue that many of the issues relating to GM crops are highly specific to particular crops and traits. The report linked this requirement for specificity to the EU regulatory regime "where individual GM crops are assessed on a case-by-case basis" (id., 27). As one illustration of this within the report, it was noted that different types of farm are likely to be affected by GM crops in different ways and the impacts of GM on the economics of agriculture could therefore be varied also. In the case of developing countries, it was suggested that "impacts will be heterogeneous and specific; GM crops are unlikely to be either a blanket solution or an unequivocal failure. Like all technologies, they may contribute to addressing certain problems, but could also exacerbate or create others" (id., 79).

In confirmation of the significant uncertainties seen to surround GM technologies, the report presents a series of future scenarios: from GM crops becoming "part of the fabric" to operating at a "bare minimum" or even "no GM." The possible costs and benefits of GM commercialization are then considered in very broad terms within these scenarios. The report concludes that trade-offs between cost and benefit will occur in any scenario, that the regulatory system will be of major importance for the control of risk and the possible co-existence of GM and non-GM crops, that public attitudes will be extremely important in determining the impact of GM crops, and that negative consumer attitudes will limit the demand for GM products considerably. The suggestion is that GM technology has tremendous future potential – provided that the regulatory system can cope. The report finishes with two links to the global:

> GM crop policy may also have wider impacts on UK science and industry. The UK has a leading position in biotechnology in Europe. The UK also has a global reputation for wider scientific excellence, and science-based industries as a whole make an important contribution to national output. This contribution, and its future expansion, are likely to be affected if changes in GM crop policy send signals, either positive or negative, about the UK's attitude to biotechnology, science and industry.   (id., 103)

> The scenarios show that the global impact of UK or EU GM policy should not be underestimated. The ability of developing countries to choose whether or not to adopt GM crop technology may be affected by considerations about the possible impact on exports to the EU. And taking a significantly different policy direction from other countries could cause serious trade tensions.   (ibid.)

So what can we suggest about the particular construction of the global that emerges within the Strategy Unit's economic analysis? Certainly, the analysis is more neutral in tone and more open in its treatment of GM than was found among the generally more skeptical public participants – as the argument for a case by case approach to commercialization underlines. Issues of the global feature prominently in this treatment – with particular attention to the global impact of any UK decision. Indeed, within this report global issues cannot be separated from the evaluation of costs and benefits since the UK economy is seen as fundamentally interconnected. Going further, economic analysis offers a global framing of the issues where specific decisions over one technology send international signals about, for example, how the UK approaches controversial technologies, and where "shocks and surprises" in quite different areas (climate change, an abrupt change in oil prices) can alter the policy context. Although global issues also featured prominently in the "public" strand, there was nevertheless a sense within the public discussion that, at least in principle, the UK could make its own independent evaluation of the ethical and political questions being raised – otherwise, why hold a national debate at all? In the economic discussion of cost and benefit, this sense of a separate track for the UK is much weaker: whatever scenario is followed, it must be seen in a global setting, as when trade tensions or reduced multinational research activity follow.

It is tempting to conclude that economics is indeed the language of globalization: presenting an individual nation as one part of a global system, seeking to make general statements across national boundaries, offering a "universalizing discourse" which brings rationality and coherence to a complex and uncertain issue. What should also be noted is that this discourse prioritizes certain issues – especially the balance of world trade and economic competitiveness – over others, although the economic significance of public attitudes is given great emphasis. Throughout, global developments and the UK scenarios are tightly coupled:

> scenarios with very limited growing or consumption of GM crops may be harder to sustain if the majority of countries in the rest of the world decide to use GM crops. Conversely, should there be a slowing or reversal in GM crop uptake globally, then a UK future with widespread GM crop growing could be problematic in the long term.
>
> (id., 95)

In the very presentation of these scenarios, there is the assumption that the UK is indeed part of a global economic system, that no national action can be insulated from the world economic order, and that there is no alternative to participation in this global system.

Strand three: The science review

At this point, I turn to a third form of discourse which offers a rather different presentation of the global: the GM science review. This expert report, led by the Government Chief Scientific Adviser, set out to summarize the "state of scientific knowledge, consensus and areas of uncertainty on GM science issues" (id., 115). In two extremely thorough reports (substantially more lengthy than those emanating from the other two strands) the main evidence and, in the second report, counter-evidence are assessed. However, the claim to globality made by the science review is very different in character to that found in the economics report:

> Science has a central role in the regulation of genetically modified organisms because it provides the evidence base for decisions on safety to human health and the environment, and it is on the basis of safety (not benefits), that approvals are granted.
>
> (GM Science Review Panel 2003, 43)

> The scientific method we currently have is by no means infallible, not least because we
> can never know "everything about everything"; but it is an approach that has stood the
> test of time, is objective and is the best method we have.   (id., 30)

The claim to globality within the science review is therefore not simply a matter of the global relevance of the issues under discussion but also bound up with the global (indeed, universal) status of the knowledge in question. Science's claim to objectivity is simultaneously a claim to speak across national boundaries but also to reflect upon the significance of those boundaries in terms of local variability. However, the review is — in some contrast to the other two reports — highly reflective on the epistemological status of the claims being made, noting the limitations to scientific understanding, the existence of uncertainties and the areas where further research is needed:

> Some of the questions asked about GM crops are purely scientific, while others are not
> of a scientific nature at all, but may be economic, social, ethical, or even personal. For
> science, as for other areas, the answers given may often depend on the way the question
> is asked and be open to different interpretations.   (id., 9)

In this way, the review can claim to have taken account of previous criticisms of the use of science in the regulatory process (Irwin 1995; Wynne 1989). The report is also keen – in line with the economics report – to argue on a case by case basis: this is not a homogeneous technology and a blanket approval (or disapproval) would be inappropriate. This suggestion from two of the strands would later form the basis of the UK government response to the GM dialogue.

What representations of the global appear within the science review? In summary form, these can be expressed in the following terms:

- The global evidence base: The report explicitly and very thoroughly brings together international evidence i.e. both evidence collected from international research teams (over thirty pages of references in the first review alone) but also such evidence as has been gathered from international experience: "Many millions of people…have for up to seven years been eating food products derived from animals fed on GM diets and no substantiated ill effects have been reported" (GM Science Review Panel 2003, 12);

- The global (but also local) applicability of results: The review presents GM as a global issue but also emphasizes contextual variability so that all countries are encouraged to conduct their own review "because the demands of agriculture, and the societies it supports, vary too much across the world to be able to reach simple generalizations" (id., 29). The knowledge claims of the review are given "global" status but contextual variability means that local circumstances must always be considered;

- The global context of GM discussion: As with the economics review, the scientific report is very aware of the potentially global application of this technology and of the concerns and questions to which it gives rise. However, the issue of GM crops in developing countries was explicitly ruled to be beyond the review's remit. Instead, the review expresses the hope that the scientific information gathered will "prove helpful to other countries" (GM Science Review Panel 2004, 12);

- The global process of science: The review presents science more generally as a thoroughly internationalized activity – a "global scientific community": "Over the past two centuries, international science has evolved a set of working principles based on the accumulation of evidence, assessment of that evidence and communication by publication, so that the global scientific community can benefit from shared knowledge" (GM Science Review Panel 2003, 30). This is not simply a question of an evidence base but is fundamental to the very status of the knowledge claims being made.

The global claim to objectivity therefore underpins the whole review even as this global construction consistently points to the limitations of its own standing – especially when dealing with questions which are not defined as "scientific" in character. Thus "science provides the evidence base for decisions on safety to human health and the environment. But it is not the basis for how effectively (scientifically achievable) regulatory requirements are enforced" (GM Science Review Panel 2004, 48). This allows the (first) report to make a link between globally inadequate diets and new GM products such as "Golden Rice" and "Golden Mustard" (which aim to increase the nutrient quality of staple crops). However, when criticized by NGOs for ignoring the social and economic causes of malnutrition, the (second) report could argue that "it simply wished to illustrate a line of further development" and that "science can only play a part in the complex issue of how worldwide malnutrition might be addressed" (id., 54).

Whilst the report on the public discussion sought to reach general statements – "People are generally uneasy about GM," "There is little support for early commercialization" – the science review attempts to be much more categorical in its judgments and self-consciously draws a line between science, on the one hand, and ethics or politics on the other. Of course, and as with the Golden Rice example, this also illustrates the problematic character of this line: "illustration" of the technical possibilities can also be understood as implicit condonation. This is an unavoidable dilemma within each of the three strands but is particularly relevant to the science review given its aspiration to objectivity. To offer a second example, there was some NGO criticism of the first report for offering an "over-optimistic picture of the safety of GM foods in the food chain by focusing on best practice, or what is theoretically possible, in GM food safety assessment rather than what is actually done" (id., 47). This drew the response that the review was concerned with what is "scientifically achievable": any "identifiable deficiency" in these "scientifically achievable aspects" was for regulatory bodies and government to consider (id., 48).

Constructing the global

Bringing together these strands of the GM dialogue, we can identify three distinct constructions of the global:

- In the public debate, there was a tendency for the global to be negatively associated with loss of national autonomy, the influence of multinationals, and the possible exploitation of developing countries. However, the minority view of more positive participants presented this global technology as having economic and social benefits – including for developing countries;
- In the economic review, globalization was characteristically understood to be a fact of economic life – the UK was seen as "embedded" in international context. Whilst global issues were taken within the public debate to be overwhelmingly ethical and political in character, the economic review presented this embeddedness in terms of international economic competition. In the public debate, GM was generally presented as a generic technology with dubious benefits. In the economics review, GM was presented as heterogeneous in form and impact. The economics review tended also to treat the technology as neutral in itself – it could be used to either positive or negative effect;
- In the science review, the global again takes different form. The review draws on a global scientific community. Somewhat paradoxically, the claims to knowledge are global in character (truth does not vary according to national boundaries) even as they also emphasize contextual variability (the truth does not vary but local contexts do). A demarcation is also established between those issues which are properly "scientific" (in the realm of objective truth) and those seen to belong to the sphere of politics or ethics. This demarcation was subsequently challenged by critics of GM (for example, in the case of Golden Rice) concerned that this demarcation might offer a misleading picture of regulatory practice. Certainly, the line between what is "scientifically achievable" and what is actual regulatory practice can be more problematic than the Science Review Panel seems to acknowledge (Irwin 2001).

One immediate question in this situation concerns the compatibility of these global perspectives, especially their compatibility with the policy process. The Government response to the debate goes to some length to emphasize the value of all three strands (especially, the public strand). In practice, however, the decision to proceed with GM technology on an "individual case by case basis" fits more easily with the economics and science strands than it does with the generally more cautious public debate. Indeed, one of the key messages from the public discussion was that there is "little support for the early commercialization of GM crops" and that people are "generally uneasy about GM crops and food." The Government response emphasizes the consistency of its decision with other applications of GM technology – such as veterinary and human medicines. This judgment would also send a global signal that the UK is open to scientific and technological innovation. However, whilst the economics and science strands are generally accepted and appear to feed directly into the Government decision, the "public" strand is presented more broadly as a matter for further deliberation and debate – an opinion which Government should bear in mind rather than a body of firm evidence on which it must act.

In illustration of this differential reception to the three strands, we can take the Government response to public concerns over the activities of globalized industry: "Although some people may be uneasy about the role of multinational companies in developing and promoting GM crop technology, we believe the appropriate response is to ensure that the technology is effectively regulated to protect the public interest" (DEFRA 2004a, 29). This sits somewhat uneasily with the public view that not just multinationals but also government is viewed with "widespread mistrust." In order to deal with this, the Government response gives great emphasis to matters of precaution, health and environmental protection, consumer choice, openness and transparency, and product labeling. Whilst the public raises fundamental ethical and political questions about globalization, Government ultimately places its faith in the global market:

> We believe we should keep an open mind and allow the technology to develop within a strict regulatory system that is designed to protect human health and the environment while providing choice. Ultimately the market will decide whether GM crops are a success or not. If farmers and consumers do not see the benefits of GM crops and GM foods, they will not grow or buy them.   (ibid.)

It would therefore appear to be the case that certain constructions of the global carry more political weight than others. This also suggests that the global can be an area of considerable contestation, where different accounts struggle for recognition – and where certain forms of global talk are more influential over the policy process than others. Rather than simply being a matter of global vs. national forces, this crucially is also an issue of who gets to speak for the global and in what form.

CONCLUSION

What does this discussion of global constructions tell us about the relationship between citizenship, scientific governance and industrially-driven innovation? In summary form, the following points emerge:

- In regulatory policy, all parties must acknowledge that "global talk" will form an important context for action and decision making rather than continuing to assume that nations can act independently. However, as we have explored, the global is not a unitary or homogeneous force. Instead, it is open to local as well as international construction and it is likely to be contested and negotiated within particular contexts;

- It seems plausible to suggest that government – and, by implication, industry – will construct globalization in a very particular fashion, generally drawing upon the language of international competition and science. Meanwhile, the UK GM case suggests that members of the public (and critical NGOs) can present the global in very different terms – as a North American invasion of national practices, as linked to organizations that are viewed in very skeptical terms, as in opposition to democratic and ethical debate;

- This discussion raises some challenging questions for democratic engagement with science and technology. If, as in the economic perspective, globalization is simply presented as a fact of life then the possibilities for individual nations to adopt a radically different course become very restricted. Even if the current applications of GM do not offer significant benefits, many governments will prefer to appear open to new innovations rather than stand outside the international economic system (and risk losing other areas of innovation and investment). However, public discussion can have the merit of at least making governmental choices explicit. Certainly, the UK Government response to the public debate took care to deal seriously with the questions raised rather than simply ignoring or dismissing them, even if, as noted above, this did not lead to broad governmental acceptance of public views;

- For industry in particular this more critical and complex account of globalization raises important issues. Whilst the economic model tends to present globalization as bringing national markets closer together through free trade and open competition, the discussion here suggests that globalization can also accentuate national/local differences and lead to critical public and consumer reactions. Although globalization can be presented as an inevitable trend, public reactions to GM suggest that it can also be a problem for multinationals – who may find consumer responses tainted by the negative connotations of the global (although interestingly this does not seem to apply uniformly across established global products and brands);

- In terms of scientific governance, one clear implication of this paper is that the relationship between national policy and global forces needs to be addressed in more explicit and open terms. It can be difficult for the governments of industrialized countries in particular to point out international constraints on local autonomy or explicitly to prioritize global industry over nationally-generated anxieties. Of course, these issues may be more sensitive in the UK, where "Europe" continues to be a divisive political issue, than elsewhere. Nevertheless, there does seem to be considerable scope for further discussion of national-global relations in the context of scientific governance and international regulatory policy;

- In more theoretical terms, this discussion has moved away from the more general, often macro-economic or political-economy based, talk of globalization as an irresistible tidal wave or as an irrevocable fact of modern life. If nothing else, the contrast between three forms of global talk allows us to see the more subtle constructions and interactions at work in speaking for the global – even if this also suggests the characteristic dominance of certain forms of global talk;

84 IRWIN

- Finally, as we saw in the GM case, contestations over the global can make a real difference to national policy outcomes – and this is likely to be an increasing trend.

Talk of the global will play an increasingly prominent role in regulatory policy and scientific governance world-wide. Decisions will need to be taken and national policies agreed. The analytical challenge throughout will be to treat the variety of forms of global talk in a critical, reflective and open fashion, recognizing always that "the global" is not simply an external force but also an expression of our underlying values and beliefs.

In practical terms, this suggests the need to move away from simply presenting globalization as an objective (and generally irresistible) force and towards an acknowledgement of its varied manifestations and social constructions. More particularly, the suggestion of this chapter is that the current hegemony of economics- and science-based responses to globalization fails to give adequate weight to matters of democratic scrutiny and public engagement. Rather than simply taking a pro- or anti-globalization stance, the political challenge is to create more open cultures of deliberation and reflection in which crucial matters of regulation and innovation are not simply presented as predetermined and beyond critical evaluation. In that way, the question of "who gets to speak for the global and in what form?" takes us to the heart of scientific governance in the contemporary world.

REFERENCES

AEBC (Agriculture and Environment Biotechnology Commission). 2001. *Crops on trial*. Available at: <http://www.aebc.gov.uk/aebc/pdf/crops.pdf> (Last accessed: 18 April 2005).
Beckett, M. 2003. Foreword by the sponsor minister. In: Cabinet Office Strategy Unit, *Field work: Weighing up the costs and benefits of GM crops*. Available at: <http://www.strategy.gov.uk/downloads/su/gm/index.htm > (Last accessed: 18 April 2005).
Browne, A. 2004. Lifting GM food sales in Europe satisfies no one. *The Times*, 20 May, 38.
Cabinet Office Strategy Unit. 2003. *Field work: Weighing up the costs and benefits of GM crops*. London: Strategy Unit.
DEFRA (Department for Environment, Food and Rural Affairs). 2004a. *The GM dialogue: Government response*. London: DEFRA.
———. 2004b. *Secretary of State Margaret Beckett's statement on GM policy*. London: DEFRA.
Friedman, T.L. 2000. *The Lexus and the olive tree: Understanding globalization*. New York: Anchor Books.
GM Nation? 2003. *The findings of the public debate*. Available at: <http://www.defra.gov.uk/environment/gm/debate/index.htm> (Last Accessed: 18 April 2005).
GM Science Review Panel. 2003. *First report*. London: Department of Trade and Industry.
———. 2004. *Second report*. London: Department of Trade and Industry.
Grove-White, R. 2002. A political con. *Spiked Online*. Available at: <http://www.spiked-online.com/Articles/00000006DA0A.htm> (Last accessed: 9 May 2005).
Hagendijk, R.P. 2004. The public understanding of science and public participation in regulated worlds. *Minerva* 42: 41-59.
Held, D., A. McGrew, D. Goldblatt, and J. Perraton. 1999. *Global transformations: Politics, economics, culture*. Cambridge: Polity.
Irwin, A. 1995. *Citizen science*. London: Routledge.
———. 2001. *Sociology and the environment*. Cambridge: Polity.

————. Forthcoming. The politics of talk: coming to terms with the "new" scientific governance. *Social Studies of Science.*

Millstone, E. 2002. The globalization of environmental and consumer protection regulation: Resources and accountability. In: *The regulation of science and technology,* ed. H. Lawton-Smith, 179-200. Basingstoke: Palgrave.

Wynne, B. 1989. Frameworks of rationality in risk management: Towards the testing of naïve sociology. In: *Environmental threats: Perception, analysis and management,* ed. J. Brown, 33-47. London: Belhaven.

Yearley, S. 1996. *Sociology, environmentalism, globalization: Reinventing the globe.* London: Sage.

CHAPTER SEVEN

THOMAS ALURED FAUNCE

GLOBAL INTELLECTUAL PROPERTY PROTECTION OF "INNOVATIVE" PHARMACEUTICALS

Challenges for Bioethics and Health Law

Multilateral and bilateral trade agreements have become important vehicles by which US multinational corporations, through close collaboration with government officials, are striving, amongst other objectives, for increasingly stringent global intellectual property protection (GIPP), particularly over what they term "innovative" pharmaceuticals.

This chapter explores the evolution and structural dynamics of GIPP. It particularly considers the hypothesis that GIPP represents a corporate-driven ideology whose legitimacy in a democratic polity is undermined by its uncertain foundation in public health research and inadequate integration with norms of bioethics and health law, including international human rights.

This analysis begins with consideration of the domestic evolution of GIPP from within the US patent system. This may reveal how many of its important structural features had their roots in a domestic profit-making ideology. The chapter then examines the critical initial globalization role of the US *Trade Act* 1974, particularly section 301. This permitted US industry to request an investigation by the US International Trade Commission of foreign nations whose practices allegedly caused it material injury. The *Agreement on Trade Related Intellectual Property Rights* (TRIPS) is analyzed as a mature component of GIPP by which increased intellectual property rights, in particular over pharmaceuticals, were linked with strong trade sanctions. The sophisticated contribution to GIPP made by the *Medicare Prescription Drug Improvement and Modernization Act* 2003 (US) is then evaluated, particularly its prohibition of Federal Government medicine price setting and its requirement for a study of pharmaceutical price controls in other developed countries.[1] In each case the extent to which GIPP attempted or failed to integrate its corporate-designed principles with basic norms of bioethics, public health law and international human rights is discussed.

B. Bennett & G.F. Tomossy (eds.), Globalization and Health: Challenges for Health Law and Bioethics, 87-107.
© 2006 *Springer. Printed in the Netherlands.*

An important GIPP case study briefly presented here involves provisions in the US-Australia Free Trade Agreement (AUSFTA) attempting to "eliminate" Australia's medicines cost-effectiveness pricing system known as the Pharmaceutical Benefits Scheme (PBS).

THE DOMESTIC ORIGINS OF GIPP IDEOLOGY

The rules and laws of intellectual property have traditionally been designed to achieve optimal balance between two ends: reward of innovation and diffusion of knowledge to the public. The philosophy underpinning them was once described as emerging from the normative traditions of natural law theory, or utilitarianism (Hettinger 1989). Such claims, as we shall see, are rarely made with any frequency or authority in relation to pharmaceuticals today.

In the late 1960s Nordhaus demonstrated that optimal duration of patent protection balanced incentives for innovation against the social losses of monopoly exploitation. He attempted to show how optimal patent life over a product is longer if price elasticity of demand is lower and its social benefit is reduced relative to research and development costs (Nordhaus 1969).

Gradually, however, such utilitarian attempts to weigh social benefit of patent monopolies, particularly over pharmaceuticals, appear to have been replaced with a more purely profit-driven economic analysis. This conceptual shift created one of the basic preconditions for the globalization of intellectual property enforcement. It is unlikely that any particular group of pharmaceutical executives devised, at one time, the complete strategy for GIPP. Rather, the hypothesis explored here is that GIPP's evolution was the outcome of a network of corporate thinking inexorably committed to maximizing profit from pharmaceuticals. The related corporate institutions, lacking institutional incentives to reward moral responsibility or ethical thinking, may have gradually eroded the confidence of ambitious individual employees to champion such non-commercial norms. The first step appears to have involved developing a strategy to favorably influence domestic economic policy, then exert greater control over the structures creating relevant US intellectual property law.

Research subsequent to Nordhaus attempted to show that optimal patent duration should be longer for economic reasons where enforcement is costly or incomplete (Scherer 1984). Likewise, the case was made that the economic incentive of patent life should be shorter where competitors wasted resources with "window dressing" inventions merely to improve market share (Gallini 1992). The traditional Nordhaus model was also contentiously modified to include what is referred to as "cumulative" or "incremental" innovation (Scotchmer 1991).

Some argued, prophetically given later globalization developments, that if such reasoning was accepted, the patent monopoly over pharmaceuticals would become a form of rent pursued by competing investors until much relevant and anticipated social benefit had been dissipated through duplication (Grady and Alexander 1992). One underemphasized line of analysis considered that much pharmaceutical innovation proceeds in public-funded institutions partly as a result of researchers' motivation to facilitate equitable dispersal of knowledge and promotion of public goods (Eisenberg 1992). Of the twenty-one drugs with greatest therapeutic effect introduced between 1965 and 1992, all but five were based on a discovery made in the public sector (Cockburn and Henderson 1997). This type of socially-focused patent law reasoning involves a strong implicit emphasis on distributive justice, a foundational principle of both bioethics and public health law (Faunce 2005). It is an approach, as we shall see, rarely engaged with as protection of pharmaceutical intellectual property became a global exercise.

In 1959 the Kefauver committee found evidence of substantial abuse of monopoly power in the US pharmaceutical industry (Comanor 1966). As a result of its recommendations, the US Food and Drug Administration (FDA) commenced a more rigorous evaluation of efficacy as well as bioequivalence in new pharmaceutical applications (Comanor 1986). Counter arguments were raised that the decline in communally valuable new drugs began well before any increase in regulatory stringency and was particularly related to tranquilizers whose supply and demand were adversely affected by the thalidomide tragedy (Temin 1980). Further, the increased regulatory requirements appeared to have no dampening effect on pharmaceutical research and development spending, which continued to rise during this period (Grabowski and Vernon 1983).

Nevertheless, the US pharmaceutical industry now promoted what was to become a common tactic in its later efforts at globalization. It blamed recently enhanced government regulation for the decreased number of innovative molecular entities it was able to introduce in subsequent years (Peltzman 1973). They argued, ultimately successfully, that FDA burdens should be relaxed and patent lives extended to compensate for market time lost in regulatory review (Wiggins 1983). Here too we see the origins of a linked accountability diversion and patent extension technique that lead to globally problematic relations between government regulators and the pharmaceutical industry.

On 9 July 1982, Barry MacTaggart, then chairman and president of the pharmaceutical company Pfizer International, published an op-ed piece in the *New York Times*. This document represented a pivotal point in the shift of pharmaceutical intellectual property protection toward a global strategy. It crystallized much initial industry thinking concerning a new target to blame for its domestic failures on the "innovation" front.

MacTaggart alleged that US knowledge and inventions were being stolen by particular foreign governments by means of specifically designed laws. The World Intellectual Property Organization (WIPO) was criticized for "trying to grab high-technology inventions for underdeveloped countries" and for contemplating treaty provisions that would "confer international legitimacy on the abrogation of patents" (Drahos and Braithwaite 2002, 61). Ominously in terms of subsequent developments, no attempt was made in this brief but seminal tract to consider how such a free market approach should mesh with exceptions to patent rights in the interest of community benefit.

Significant in the domestic background of this influential public enunciation of GIPP was the 1980 decision of the US Supreme Court in *Dawson Chemical Company v. Rohm and Haas.*[2] This overruled prior decisions where judges deprecated patents as disguised, socially disadvantageous monopoly rights. The Court now declared that "the policy of free competition runs deep in our law…but [that] of stimulating invention…underlies the entire patent system [and] runs no less deep." In *Haas,* reward of "innovation" though State grant of protectionist monopoly rights achieved the status of "equal footing" with the previously antagonistic concept of "free market competition" (Kastriner 1991, 7).

In 1982, another important element in GIPP arose from the creation of the Court of Appeals for the Federal Circuit (CAFC). This Court's ostensible purpose was to centralize patents, tariff and custom, technology transfer, trademarks, government contracts and labor disputes within one specialist jurisdiction. Critics feared the new court would be prone to isolation from broader normative systems and to influence by corporate interest groups (Lever 1982). Yet these were probably two of the main reasons for its creation. For GIPP to begin to launch itself upon the world it first required a solid and consistent basis in domestic patent law, one that unequivocally emphasized the paramount importance of the rights and profits of the innovator.

The CAFC has since, as expected, developed an extremely pro-patent jurisprudence rarely mentioning the word "monopoly," readily granting large scale compensatory damages and permanent injunctions, whilst consistently upholding the interests of alleged innovators over purported copiers or generic suppliers. The adverse social impacts of such decisions, and the extent to which they conflict with basic principles of bioethics or public health law are rarely, if ever, discussed in this new patent court jurisprudence (Sell 2003, 67-72). Between 1982 and 1990, the CAFC upheld on appeal 90 percent of patents initially determined to be valid and infringed, compared with 62 percent in the various relevant courts between 1953 and 1978. It reversed on appeal only 28 percent of patents held invalid at first instance, compared with 12 percent previously (Jaffe 2000). The CAFC later produced many decisions that appeared very advantageous to the development of GIPP. Also assisting the nascent GIPP ideology was industry lobbying for a Federal economic policy positing level of output, rather than amount of competition as the dominant regulatory end point. This allowed pharmaceutical companies in particular to promote high levels of market concentration as efficiencies, rather than price-distorting monopolies and cartels that conflicted with ethical and legal obligations to promote competitively low prices in the public interest (Sell 1998). GIPP was beginning to emphasize that intellectual property rights provided such corporations with a strategy to protect investments and increase revenue, if need be by excluding competition from the market. This was quite different from earlier conceptions, which stressed the role of patents in social diffusion of knowledge (Sell 2003, 13-4).

In 1983 the US Government passed the *Drug Price Competition and Patent Restoration Act* (commonly known as the *Hatch-Waxman Act*). This legislation gave pharmaceutical patent holders an additional five years of patent life, allegedly to compensate for the period of pre-market testing and FDA evaluation. It allowed, as a response to the CAFC decision in *Roche Products, Inc. v. Bolar Pharmaceutical Co,*[3] generic competitors to use original brand-name data to prepare bioequivalence and other testing provided those activities were reasonably related to securing regulatory approval and "springboarding" on originator patent expiry. The statute provided an incentive for rapid generic market entry by according the first such entrant 180 days of market exclusivity. Brand name manufacturers were allowed to request a thirty-month injunction against marketing approval of generic drugs alleged to be infringing valid patents (Gallini 2002). This last provision in particular, by linking marketing approval (previously based primarily on safety and quality issues) with patent validity, established the profitable practice, known as "evergreening," by which the patent monopoly over large sales volume brand name pharmaceuticals could be tactically extended. The techniques of "evergreening" developed here were set to be transported, by their incorporation into bilateral trade deals, into public health systems around the world as a key component of GIPP.

The next important development in GIPP ideology appears to have involved the search for an effective mechanism to widen the global markets over which pharmaceutical intellectual property rights could be enforced (Sell 2003, 17). To do this, US pharmaceutical companies successfully prosecuted the argument that any form of State restriction on their prices in foreign countries was an unjustified interference in the marketplace, rather than an ethically and legally legitimate public health restraint on a protectionist market distortion (Drahos and Braithwaite 2002, 13). This again was largely an ideological debate. Very little objective evidence of public health impact was adduced either for or against GIPP.

Section 301 of the *Trade Act* 1974 (US) provided GIPP with an initial, largely unsuccessful, global enforcement mechanism. Between 1975 and 1979, eighteen corporate petitioners filed section 301 cases, but in none managed to induce the US to take retaliatory action, six being settled by bilateral resolution (Coffield 1981). Section 301 of *Trade Act* 1974 (US) was amended in 1984 to permit the US President, through the office of the US Trade Representative (USTR), to deny trade benefits or impose duties on products or services of countries unjustifiably restricting US commerce. In 1987, the US Pharmaceutical Research and Manufacturers Association or PhRMA (then called the PMA, or US Pharmaceutical Manufacturers Association), demanded trade retaliation against Brazil under section 301, for the latter's lack of adequate patent protection for US pharmaceuticals. No serious attempt was made by the US to balance the ethical and public health law obligations of the Brazilian government to provide affordable, essential medicines for its citizens. When Brazil refused on social justice grounds to alter its policy, the US placed a large retaliatory tariff on imports of Brazilian pharmaceuticals. Brazil filed a complaint with the GATT, but withdrew this when US sanctions were dropped in return for a commitment to increased pharmaceutical patent protection (Mossinghoff 1991).

In 1987, the US also denied trade benefits to Mexico because of that country's failure to adequately protect US pharmaceutical patents. The Mexican government refused to buckle to this pressure, holding instead to a longstanding public health commitment to provide affordable, essential medicines to its people. This persistent refusal lost Mexico $500 million in Generalised System of Preferences (GSP) benefits (Sell 2003, 90-1). The GSP scheme, set up under the United Nations Conference on Trade and Development provided for preferential tariff treatment for developing country exports of manufacturing and semi-manufactured goods. In 1988, an amendment called "Special 301" was made to section 182 of the *Trade Act 1974* (US) by the *Omnibus Trade and Competitiveness Act 1988* (US). Special 301 became the principal statutory authority under which the US investigated and, if need be, threatened trade sanctions against foreign countries that maintained acts, policies and practices that violated, or denied US corporations rights or benefits under trade agreements, or, through otherwise being unjustifiable, unreasonable or discriminatory, burdened or restricted US commerce. Unjustifiable acts, policies and practices were defined as those that violated, or were inconsistent with, the international legal rights of the US, including denial of national treatment or most-favored nation (MFN) treatment to US exports or limiting protection of US intellectual property rights.

The USTR was now required under the *Trade Act* 1974 (US), to mention, in its annual review, a "Special 301 Report Priority Watch List." Corporations could petition the USTR to investigate and, ultimately, threaten trade sanctions against a particular unjustifiable, unreasonable or discriminatory policy or practice of a foreign country so listed. The Special 301 Report Watch List will probably remain a classic example of public law at the service of private corporations (Drahos and Braithwaite 2002, 89). The 2004 list includes, for example, these comments about Canada's attempts to restrain pharmaceutical prices for social justice reasons in the public interest: "systemic inadequacies in Canadian administrative and judicial procedures continue to allow the early and often infringing entry of generic versions of patented medicines into the marketplace" (USTR 2004).

Croatia is likewise criticized because of its "lack of co-ordination between the patent and health authorities to prevent patent infringement by the grant of marketing approval for copycat pharmaceuticals, and failure to provide expeditious and timely judicial remedies to parties seeking to stop infringing activities" (ibid.). Ecuador is similarly impugned because "the number of copy products granted marketing approval by the health authority continues to increase, due to the lack of any linkage system between the health and patent agencies" (ibid.).

The policies of the Italian government were attacked on the USTR "Priority Watch List" at US multinational corporate insistence. The USTR reasoned that the policies "may adversely affect the prior practice of patent term extension for pharmaceuticals" (ibid.). Malaysia is denigrated for failing to link "the marketing approval process to the patent registration process" for pharmaceutical products (ibid.). Poland is criticized for permitting the commercial availability "of generic versions of patent protected pharmaceutical products" (ibid.). Vietnam is castigated because "counterfeit pharmaceuticals are common in the marketplace" (ibid.).

The USTR Priority Watch List consistently avoids opportunities to consider the human cost of the more stringent patent protections they so stridently advocate as a global legislative priority. No attempt, for example, is made in any such Special 301 report, to balance the public health cost of requiring the impoverished citizens of these mostly developing nations to pay higher prices for pharmaceuticals. The available income, or burden of disease in these countries is not mentioned as a relevant factor. No effort is made here to address obligations flowing from bioethics and public health law (particularly distributive justice) or international human rights (the right to health). The next stage in GIPP's evolution involved linkage to an even stronger method of global intellectual property enforcement.

GIPP'S LINKAGE WITH TRADE

From 1981, Edmund Pratt, then CEO of the Pfizer pharmaceutical company, in his capacity as chair of the Advisory Committee on Trade Negotiations (ACTN) had been consulting directly with the US President about placing foreign intellectual property protection on the US trade agenda (Ryan 1998). At this time, the US commenced a series of bilateral negotiations on patents, copyright and trade with countries such as Korea, Mexico, Singapore, Hungary and Taiwan. US intellectual property negotiators apparently discovered, however, that financially more effective outcomes emerged once their trade colleagues did most of the bargaining (Enyart 1990, 54).

The task of making GIPP a primary object of US trade policy was skillfully executed. Pharmaceutical company lobbyists, as mentioned earlier, had previously sowed the idea of linking trade and intellectual property rights in various levels of the relevant US bureaucracy, government and academia. At the same time, they increased the size of their contributions to the election campaign funds of the two major US political parties. The World Intellectual Property Organization (WIPO) seems to have been by-passed in the task of promoting GIPP, perhaps because the US and other OECD nations considered it lacked sufficient enforcement tools or motivation (Abbott 2002, 315).

In the mid to late 1980s, when GIPP was increasingly being expressed in TRIPS negotiations, very little empirical or theoretical research existed concerning the effects of increasing intellectual property rights in a country (id., 313). Some evidence had emerged that stronger patent rights appeared to encourage incremental improvements by the originator, but also to create hindrance from prior inventors and freeze out future radical inventors (O'Donoghue, Scotchmer and Thisse 1998). Lerner (2001), for example, studying 177 policy shifts in 60 countries over 150 years, found an "inverted-U" relationship between patent strength and innovation. He suggested that strengthening patents had a positive effect on innovation when intellectual property protection was initially low, but a negative impact if patent protection was initially high (ibid.). The limited research that did support the trade-IP linkage with pharmaceutical innovation was generally written by drug company-funded institutes and academics (Bekelman and Gross 2003). This too appears to have become a consistent tactic of GIPP proponents. When counterarguments to GIPP are raised by academics or policy makers, a paper is rapidly published allegedly either confirming their lack of economic rationality, or supporting the GIPP position.

The TRIPS Agreement was a manifestation of GIPP developed by senior executives at twelve US corporations including, in particular, the pharmaceutical giant Pfizer (Drahos and Braithwaite 2002, 61). Its standards were designed to obtain rent for developed nations from two great emerging technologies, digital technology (through copyright, patents and protection for layout designs) and biotechnology (through patents and trade secrets) (id., 10). Developing countries, led by India and Brazil, both of whom had large generic drugs industries catering to the essential needs of impoverished populations, fought against TRIPS in the 1986 Uruguay Round mandate of the WTO. Amongst their objections were suspicions that increased intellectual property protection would burden the task of providing universal access to affordable, essential medicines. The US negotiators bargained that they at least be allowed to place the issue "on the table." Eventually, a Senior Officials meeting was convened in Geneva in April 1989 that announced a "framework text" to provide the basis of substantive negotiations whilst not considering issues of institutional implementation (id., 11).

The US, European Community and Japan, however, signified that subsequent economic co-operation between their nations and the developing world was dependent on TRIPS Agreement being reached. The US negotiators made quite open threats about trade sanctions under Special 301 of the *Trade Act* 1974 (US), or even abandoning GATT altogether (id., 193). Developing countries were promised the developed parties to TRIPS would strive to reduce their domestic agricultural subsidies and alleviate restrictions on the import of tropical products. Although it was clear that TRIPS would have a major impact on public health particularly in developing countries, the World Health Organization was not included in the negotiations (ibid.).

The developing countries, led by India, strove to achieve explicit concessions under TRIPS to allow liberal use of compulsory pharmaceutical licensing. They also sought to specifically reduce restrictions on the parallel importation of drugs from countries where the price was cheaper, as a result, for example, of national bargaining strategies (Abbott 1998, 500). Neither goal was fully achieved; Article 30 instead conferring an ambiguous general right of "limited exceptions to the exclusive rights conferred by a patent." Article 7 recognized that the protection of intellectual property should contribute to the promotion of technological innovation and to the transfer and dissemination of technology, to the mutual advantage of its users and producers in a manner conducive to social and economic welfare and to a balance of rights and obligations. A clarification would be needed, however, to determine the extent to which this Article allowed public health exceptions to pharmaceutical patent rights.

Developing nations agreeing to TRIPS were asked to renounce a large portion of their sovereignty over areas of intellectual property that would be crucial to ensuring justice and fairness dominated the policies that shaped their subsequent economic progress. The first reason they did so, nevertheless, may have arisen from the belief that this would accord them greater agricultural access to developed world markets. Ironically, the developing nations actually obtained little such advantage from TRIPS in the agricultural sector. Instead, the TRIPS regime increased developed nation intellectual property rights over agricultural plants and seeds, facilitated direct competition with biotechnologically superior developed nation farmers, allowed US and EU farmers to continue benefiting from large farm subsidies and permitted developed nations to use phytosanitary restrictions as a de-facto means of protection against developing nation agricultural imports (Drahos and Braithwaite 2002, 11).

The second reason developing nations may have agreed to TRIPS is that they were somehow convinced that higher intellectual property standards would eventually benefit them. Yet, then and now, developing nations hold a minute proportion of the world's patents. Further, industrialization had began to occur in Singapore, Brazil, India and South Korea well in advance of a globalized intellectual property regime (id., 143).

TRIPS eventually emerged as one of the twenty-eight agreements in the *Final Act of the Uruguay Round of Multilateral Trade Negotiations* leading to the WTO in 1994. It required all signature countries to adhere to minimum levels of intellectual property protection (including pharmaceutical patents). It was the first broadly subscribed multilateral IP agreement enforceable between governments and allowed them to resolve international intellectual property disputes more readily through the WTO dispute mechanism. Developed countries were required to fully implement TRIPS by 1 January 1996, while developing and least developed countries were given staggered compliance periods (Gathi 2002).

TRIPS and the GIPP ideology it appeared to implement, may historically be viewed as a triumph of corporate lobbying over democratic bargaining. "A small number of US companies, which were established players in the knowledge game, captured the US trade-agenda-setting process, and then, in partnership with European and Japanese multinationals, drafted intellectual property principles that became the blueprint for TRIPS" (Drahos and Braithwaite 2002, 11).

In 1997 the South African government, as the result of the HIV/AIDS crisis affecting 50 percent of its citizens in some districts and its inability to respond with cheap anti-retroviral medications, passed its *Medicines and Related Substances Control Amendment Act.* Section 15C permitted the relevant Minister to "prescribe conditions for the supply of more affordable medicines in certain circumstances so as to protect the health of the public." It expanded the conditions for compulsory licenses and parallel importation to facilitate the capacity of more poor South African citizens gaining access to cheap anti-HIV/AIDS pharmaceuticals.

As a result, South Africa was placed on the USTR Priority Watch List. The US Department of State, Department of Commerce, Patent and Trademark Office, USTR, National Security Council and Office of the Vice President commenced an assiduous, concerted campaign to persuade the Government of South Africa to withdraw or modify aspects of section 15C which were considered inconsistent with its commitments under TRIPS (USTR 1999).

The US threatened to bring the South African legislation before a WTO Dispute Settlement Body. The US (and the PhRMA interests it represented) claimed that the South African public health and equity interpretations of TRIPS compulsory licensing and parallel importation articles were inconsistent with TRIPS. As one commentator pointed out, however, such a view is repugnant to basic principles of moral responsibility:

> One would truly have to lack a moral compass to render a legal opinion condemning millions of people to a premature death because Pfizer, Pharmacia & Upjohn, Glaxo SmithKline or Novartis would not be able to engage in their optimal pricing strategy. The WTO, in my view, could not survive such a decision.   (Abbott 2002, 321)

Nevertheless, in 1998, as is now well known, forty-one pharmaceutical companies commenced litigation, partly based on TRIPS, against the South African *Medicines and Related Substances Control Amendment Act*. President Nelson Mandela was named as first defendant. In April 2001, the action was withdrawn after a campaign by members of the international civil society including Médecins Sans Frontières, relying in particular on basic norms of bioethics and the international right to health. Other nations (for example Brazil and Venezuela) brought and won similar cases based on their constitutional rights to health (Drahos and Braithwaite 2002, 6). These conflicts were the inevitable result of GIPP's prior refusal to engage with norms of bioethics, public health law, or international human rights.

Partly in response to lobbying by such NGOs at the WTO Doha Ministerial Conference in November 2001,[4] WTO Ministers issued a separate *Declaration on the TRIPS Agreement and Public Health*.[5] Paragraph 6 of this equity clarification permitted WTO Members with "insufficient or no manufacturing capacities in the pharmaceutical sector" to issue compulsory licenses for the production, or importation, of medicines without consent from the patent holder, where necessary to protect public health (such as to combat the HIV/AIDS crisis) and promote "access to medicines for all." After a further WTO decision of 30 August 2003, Members could unequivocally waive Article 31(f) of TRIPS and respond to compulsory licenses to export to markets (other than domestic ones) where that other country does not have the capacity to manufacture medicines itself. This practical extension of the international human right to health was not restricted to situations of national emergency. The Ministerial Declaration stressed the importance of "implementing and interpreting" TRIPS "in a manner supportive of public health, by promoting both access to existing medicines and research and development into new medicines." But if the *Doha Declaration* appeared to provide an obstacle to GIPP, this was only temporary.

At the start of the twenty-first century GIPP continued to be fuelled by the lavish expenditure of US pharmaceutical companies on marketing, administration and lobbying. In the fiscal year July 2003-June 2004, the US drug industry (brand-name, generic and biotech drug makers, biomedical device makers, pharmacy benefit managers and distributors) spent US$108.6 million chiefly on funding 824 lobbyists (many of whom had previously worked for government) to influence public policy (in 2002 $91.4 million for 675 lobbyists) (Public Citizen 2004). The CAFC decision in *Madey v. Duke University*[6] was also very useful to GIPP proponents. They were increasingly attempting to argue that pharmaceutical research and development leading to product innovation could only consistently arise not due to public-funded efforts at universities, but from corporate profit motive in a context of strong IP protection. The CAFC held in *Madey* that any experimentation into a pharmaceutical compound at a university potentially breaches a patent, because these institutions are profit-making.[7] The decision was unsuccessfully appealed to the US Supreme Court by the Association of American Medical Colleges, the American Council on Education, various individual colleges, universities and medical schools, as being contrary to basic ethical principles and inhibiting research into socially important but unprofitable diseases such as malaria, tuberculosis, diarrhoea and pneumonia.[8]

Another example of the way GIPP began to influence regulatory structures concerns its close involvement with the US FDA. A structural conflict of interest increasingly kept FDA safety officials from strongly exercising independent authority because the Office of Drug Safety is part of the section responsible for evaluating and approving new drugs. Many of its own employees consider the FDA now views the drug industry as a financially supportive client to be appeased rather than a potential infringer to be carefully regulated. In an internal survey conducted in 2002 of about 400 FDA scientists, two-thirds said they lacked confidence that the FDA "adequately monitors the safety of prescription drugs once they are on the market" and 18 percent reported they "have been pressured to approve or recommend approval" for a drug "despite reservations about the safety, efficacy, or quality" (Fontanarosa, Drummond and De Angelis 2002, 2648).

The *Medicare Prescription Drug Improvement and Modernization Act* 2003 (US) was a significant domestic triumph for GIPP. It had begun as a measure to assist senior citizens and people with disabilities cope with rising US health care costs. It ended up instead a boon for the pharmaceutical industry and managed care corporations. The US government was specifically prohibited from using its bulk buying power for Medicare beneficiaries (as the Australian government does under the PBS) from negotiating medicines price discounts.[9] This provided the stable conceptual base from which US corporations could begin a concerted program to dismantle reference pricing systems around the world. The 180-day period of market exclusivity designed to be an incentive for generic drug market entry was abolished.[10] Brand name drug companies, by injuncting the first such generic for potentially breaching their claimed patents, had effectively been prolonging their maximum profits over blockbuster (high sales volume) brand name medicines. The legislation commissions a study but fails to make legal the reimportation of prescription drugs from Canada and other industrialized countries where they are approximately half US prices.[11]

The legislation also directed the Secretary of Commerce, in consultation with the International Trade Commission, the Secretary of Health and Human Services and the United States Trade Representative, to conduct a study and report on drug pricing practices of countries that are members of the Organization for Economic

Cooperation and Development and whether those practices utilize nontariff barriers with respect to trade in pharmaceuticals. The study was required to include an analysis of the use of price controls, reference pricing, and other actions that affect the market access of United States pharmaceutical products. It was also to estimate "additional costs to U.S. consumers because of such price controls and other such practices, and the extent to which additional costs would be reduced for U.S. consumers if price controls and other such practices are reduced or eliminated."[12] Pharmaceutical price controls in eleven OECD countries were eventually studied. The resultant report by the US Department of Commerce is replete with classic expressions of GIPP ideology. Increased intellectual property protection is justified here as a necessary prerequisite to ensuring innovative new drugs. US citizens, the study claims, will gain $5 to $7 billion per year in benefits from new drugs if the OECD countries had no price controls. The report sets the benchmark for pharmaceutical prices as that in the US, because it allegedly represents a completely "deregulated" market (US Department of Commerce 2004, 7).

A Conference Agreement on the legislation obliged the United States Trade Representative, the Secretary of Commerce, and the Secretary of Health and Human Services to analyze

> whether bilateral or multilateral trade or other negotiations present an opportunity to address these price controls and other such practices and shall develop a strategy to address such issues in appropriate negotiations. In so doing, these agencies shall bear in mind the negotiating objective set forth in the Bipartisan Trade Promotion Authority Act of 2002 to achieve the elimination of government measures such as price controls and reference pricing which deny full market access for United States products. In so doing, the agencies shall provide periodic and timely briefings for the Committees of the House and Senate listed above, with an interim briefing no later than 90 days after enactment to address negotiations to establish a U.S.-Australia Free Trade Agreement and, as appropriate, other current negotiations.[13]

GIPP'S SWITCH TO BILATERALS AND THE AUSFTA

Frustrated by public health inroads into the lucrative TRIPS pharmaceutical patent arena, such as the *Doha Declaration on TRIPS and Public Health,* US multi-nationals now began to forum shift in an effort to promote GIPP. Their strategy was to negotiate tougher intellectual property regimes outside TRIPS, in private Free Trade Agreements ("FTAs"). The AUSFTA which entered into force on 1 January 2005, was in fact merely one of a series of what may more accurately be termed bilateral corporate colonization arrangements (or preferential trade agreements) negotiated with thirty-four countries in the Free Trade Area of the Americas Agreement, five Central American countries, the Dominican Republic, the Southern African Customs Union, Morocco, Bahrain and Singapore.

These FTAs had the following common features that represented core elements of the now mature GIPP. First, governments signing them were required to extend pharmaceutical patent protection beyond the twenty-year period required by TRIPS. Second, compulsory licensing, rare in any event, was nonetheless expressly limited, unlike TRIPS, to situations such as "national emergencies of extreme urgency." Third, restrictions were imposed on the parallel importation of cheap medicines. Fourth, generic companies were restricted in their capacity (allowed under TRIPS) to "springboard" by using brand name data. Finally brand name "evergreening" provisions were introduced that linked pharmaceutical marketing approval for generic medicines with notification to relevant brand name manufacturers and an assessment of their existing patent validity (Oxfam International 2004).

With the AUSFTA, however, GIPP achieved a unique breakthrough. For the first time provisions were included in a bilateral trade deal aimed at facilitating the "elimination of government measures such as price controls and reference pricing which deny full market access for United States [pharmaceutical] products."[14] The particular reference pricing system targeted here was Australia's PBS. In a now well-rehearsed tactic, Australian negotiators were told that even the limited promised access to the US manufacturing and agricultural markets would be closed unless the PBS was part of the deal (Drahos and Henry 2004).

The Deputy US Trade Representative (USTR), for example, stated that the AUSFTA was the first US bilateral free trade deal that included:

> special provisions addressing market access for pharmaceuticals...[which, to address] Australia will [have to] make a number of improvements to its Pharmaceutical Benefits Scheme...the [AUSFTA] also establishes a Medicines Working Group that will provide a forum for ongoing dialogue on Australia's system of comparing generics to innovative medicines.   (Shiner 2004, 2)

PhRMA through its representatives on a committee called IFAC3, worked closely with US trade negotiators to insert strategically planned articles in the AUSFTA (Drahos et al. 2004, 2). The close industry-government relationship which had been so important in the evolution of GIPP, was evidenced here by the fact that a former Australian government staffer had become the Chief Executive Officer of the Australian version of PhRMA, Medicines Australia (Metherell 2002). Simultaneously, a senior advisor to the former Australian Federal Health Minister, had become a PhRMA advisor on AUSFTA strategies in relation to the PBS (Beaumont 2004).

The GIPP attempt to eliminate PBS reference pricing through the AUSFTA had three major overt strategies. First, under the heading of "transparency" Annex 2C incorporated provisions requiring (under the threat of Chapter 21 trade sanctions) that PBS procedures make it more difficult to refuse to "list" any new pharmaceutical. It created an "independent review" process for decisions of the Pharmaceutical Benefits Advisory Committee (PBAC) not to list submitted medicines. It also increased opportunities for lobbying of PBAC members and established a Medicines Working Group tasked with developing procedures related to the above transparency requirements (Drahos et al. 2004).

Second, the intellectual property chapter (Chapter 17) included provisions facilitating brand name drug patent "evergreening" (Article 17.10.4). This provision required that generic drug market entry be notified to a brand name manufacturer and then indefinitely "prevented" whenever a patent was "claimed" over a brand name drug. Implementing Australian legislation created a notification process, but also imposed damages when brand name manufacturers used this process to "evergreen" their patents over blockbuster pharmaceuticals. Chapter 17 of the AUSFTA included articles that restricted the capacity of the Australian government to compulsorily license medicines in public health emergencies (Article 17.9.8), export them to deal with such crises in neighboring countries lacking their own manufacturing capacity (Article 17.9.6), or parallel import its cheaper medicines back to the vast US market to the benefit of US citizens (Article 17.9.4). Other provisions locked Australia into expanded brand name patent terms, for example in situations of delayed marketing approval that again went beyond what was required under TRIPS (Article 17.9.8).

Perhaps of most concern for the long-term viability of PBS reference pricing were the interpretive principles at the commencement of Annex 2C. These laid bare some of the core principles of GIPP. Three of them emphasized reward of "innovation" in pharmaceutical development and the fourth stressed an expectation that the respective governments would "recognize" research and development in this area. These principles are more important than might first be thought. The chief strategy employed appears to have involved shifting the emphasis of the PBS from the democratically legitimate norm (having specific constitutional and judicial support) supporting universal access to affordable, essential medicines, toward what may be termed a "corporate lobbying principle" requiring State recognition of pharmaceutical innovation and research and development. The expression "corporate lobbying principle" refers to a norm that, despite its incorporation in a trade deal and subsequent avid promotion amongst bureaucracy and government, has no established basis in systems such as bioethics, public health law, or international human rights. "Recognition of pharmaceutical innovation," "patent-driven research and development," "transparency" and "market access," for example, struggle to satisfy a legal rule of recognition in representative democracies. Bilateral trade deals are one of the most effective means by which such principles can insinuate themselves within the recognized decision making systems and regulatory structures of a democracy and alter the thrust of their respective activities toward facilitating the corporate agenda of maximizing profits and reducing costs.

The operation of such "corporate lobbying principles" in Annex 2C is facilitated by the presence in the AUSFTA of a non-violation nullification of benefits article that specifically covers both Chapter 2 (containing Annex 2C on pharmaceuticals) and Chapter 17 (on intellectual property). This provision, Article 21.2(c), permits dispute resolution proceedings to be initiated where the legitimate expectations of a party (as established, for example, by the interpretive principles at the commencement of Annex 2C) have not been fulfilled. Prioritizing reward of innovation and research and development is a dominant goal of GIPP. It is not a goal of the PBS, which instead has the task under section 101 of the *National Health Act 1958* (Cth) of comparing new medicines against existing ones in that class so as to facilitate government assistance in pricing only for those drugs that are objectively established as having therapeutic benefit.

Comments made before and after the AUSFTA negotiations indicated that the US saw the PBS strategy in this agreement as a valuable precedent for the implementation of GIPP and elimination of medicines reference pricing systems elsewhere in the world (US Department of Commerce 2004, 7). Australian determination to protect the PBS variant of reference pricing may be an act of global health significance.

GIPP, BIOETHICS AND PUBLIC HEALTH LAW

Throughout this chapter it has been emphasized that one of the most distinctive features the GIPP ideology is its lack of engagement with values and principles of bioethics or public health law. One frequent example, ignored by GIPP, has been distributive justice. John Rawls' philosophy has placed the social virtue of distributive justice at the foundation of those principles and rules which evolve in a democratic legal system to equalize, as far as possible, the basic conditions of life available to each of its citizens. Distributive justice and the international right to health require a society to attempt to reduce the burden of illness in its population and, at this point in time, to do so by ensuring that those citizens have equitable access to affordable, essential medicines (Ruger 2004). It is likely to find expression in a new social responsibility provision of the UNESCO *Universal Declaration on Bioethics* (Berlinguer 2004).

The human right to health has become an important feature of constitutional and international human rights jurisprudence (Leary 1994). At its core, expressed in the *Universal Declaration of Human Rights* (Article 25(1)) and *the International Covenant on Economic, Social and Cultural Rights* (Article 12(1)), as well as other international conventions and constitutional provisions, it obliges a State to make effective use of available resources to at least progressively realize its capacity to fulfill public health responsibilities, including the basic preconditions for health (Toebes 1999). Universal access to affordable, essential medicines is arguably now a core element of the international right to health.

Organizations such as Médecins Sans Frontières view equitable provision of essential medicines to the developing world as a core component of their commitment to concepts of distributive justice and international human rights (Médecins Sans Frontières 2005). They state that what they have witnessed since the inception of TRIPS is not an improvement in access in this area, but a deterioration.

Basic principles of bioethics and health law, that should have been taken into account by documents expressing the GIPP ideology, underpinned many core recommendations of the Millennium Summit in September 2000. The four particularly relevant targets were:

- Target Five, by 2015, reduce by two-thirds the mortality rate among children under five;
- Target Six, by 2015, reduce by three-quarters the maternal mortality ratio;
- Target Seven, by 2015 halt and begin to reverse the spread of HIV/AIDS; and

- Target Eight, by 2015, halt and begin to reverse the incidence of malaria and other major diseases (Singer and Gregg 2004, 25).

In 2000 the World Health Organization (WHO) estimated that one third of the world's population lacked access to essential drugs, with this figure rising to 50 percent in the poorest parts of Africa and Asia (WHO 2000). According to fundamental principles of bioethics and the international right to health, WHO's Model List of Essential Drugs (those which satisfy the core health care needs of the majority of the population), should be affordable and represent the best balance of quality, safety, efficacy and cost for a given health setting (WHO 2000). No core document of GIPP, however, has seriously addressed such targets or recommendations.

Instead, the proponents of GIPP promote attempts to win public relations kudos by expressing dismay at the public health crises in developing nations, while allowing GIPP to proceed unhindered. One particular example was a paper that allegedly discovered that in sixty-five developing nations covering a population of four billion, patenting is rare for 319 pharmaceuticals on the World Health Organization's Model List of Essential Medicines. This conclusion was said to justify a more "pragmatic" approach, so that public health policy might concentrate instead on "greater causes of epidemic mortality, which now pose unprecedented threats to global peace and security" (Attaran 2004, 157). Academics generally supportive of PhRMA claimed this data "should squash once and for all demands for compulsory licensing of patented medicines in most poor countries and make largely irrelevant the *Doha Declaration on TRIPS and Public Health* (Bate 2005).

Yet, in South Africa, every three-drug antiretroviral (ARV) cocktail is blocked by patents and most ARV drugs in South Africa are patented. There are four to five million HIV positive persons in South Africa and the economy there has more than 40 percent of the GDP for sub-Saharan Africa. Third, entry into the South Africa market is necessary for generic suppliers to reach the economies of scale (volume) needed for the efficient production.

Studies on the effect of GIPP throughout the period of its evolution consistently suggest it has a deleterious effect on public health. It appears that GIPP has had little positive impact on the level of local medicinal research and development whilst substantially increasing medicines prices in the countries forced to introduce it (Nogues 1990). Kawaura and La Croix showed that GIPP's introduction to Korea substantially reduced that country's wealth (La Croix and Kawaura 1996). Conclusions about the impact of GIPP in developed nations such as Japan (Sakakihara and Bransteeter 1988) and Canada (Pazderka 1999) depend on the extent to which generic pharmaceutical competition is permitted to continue, or be fought for. The evidence is weak that increased patent protection will stimulate additional R&D expenditure in countries introducing it (Deardorff 1992). The welfare costs in India may be substantial (Chaduri, Goldberg and Panle 1993). It appears likely, should it continue, that GIPP will lead to at least a 25 percent increase on global spending on patented drugs, even if China is not taken into account (Lanjouw 1997).

Those US academics whose work supports aspects of what has been described here as GIPP, are now well known (CP-Tech 2005). One of the central components of GIPP ideology is that increased pharmaceutical patent protection and prices are justified because of the high research and development costs and vast promised benefits to health care. For example, DiMasi, Hansen and Grabowski provide (2003) a figure of $800 million for the total pre-marketing approval cost for each new drug, from work performed at the Tufts Center for the Study of Drug Development (DiMasi, Hansen and Grabowski 2003). PhRMA, under the banner "New Medicines, New Hope," cites these results to claim that pharmaceutical companies rely on government-granted patents to protect their huge investments in researching and developing new drugs (PhRMA 2005). On the other hand, the US National Institutes of Health financed TB Alliance Report, *The Economics of TB Drug Development*, put the cost at between $115 to $240 million, including costs of failures (Global Alliance 2001). Others claimed that the Tufts Center calculations failed to adequately account for the substantial contribution of research from public institutions, the government subsidies and tax subsidies and the disproportionate profits of the industry. Further, the figures made controversial assumptions about capital costs, were averaged and did not accurately relate to specific drugs as no such data exists (Corea 2001, 3).

CONCLUSION

For several decades now, US trade policy has oscillated between multilateral agreements, such as the WTO TRIPS Agreement, and bilateral trade deals, in the pursuit of preferentially protectionist ideological objectives. A key component of these objectives has been the strengthening of global intellectual property protection over allegedly innovative pharmaceuticals owned by US corporations. Critical to this strategy has been the capacity to threaten trade sanctions against GIPP non-complying nations.

By the end of 2005, over 300 bilateral free trade agreements will have been notified to the WTO. They are despised by many respected economists as mechanisms whereby small countries are pressured into accepting special interest provisions of GIPP that actually besmirch the name of "free trade" (Irwin 2005). Such deals appear to be turning the world trading regime into a "dog's breakfast" of constructive ambiguities and preferentially protectionist intellectual property rules, particularly advancing the corporate agendas of pharmaceutical multinationals. It is a tactic that fragments the coalitions of developing countries on core social justice issues in multilateral trade negotiations, to the public health detriment of their citizens.

One hypothesis is that for the US government to have so aggressively sought to implement GIPP, it must have fundamentally shifted its commitment to normative systems which advocate primacy in policy for the concepts of respect for human dignity and the egalitarian relief of human suffering. These systems, bioethics, public health law and international human rights, have not been engaged with by GIPP, restricting its credibility and legitimacy.

Indeed, a significant, emerging problem for global public health is that the increasingly dominant GIPP ideology has developed no accepted, intrinsic mechanism for weighing considerations of overall community benefit against its desire to maximize profit. Its central articles of faith continue to propound the ideology that anything interfering with free markets makes them less efficient, and that people are best conceived as suppliers and demanders of commodities at the service of manufacturers and investors. Gestures are periodically made and marketed to show the compassionate side of pharmaceutical multinationals (Wehrwein 2002). Yet, carefully orchestrated displays of strategic pharmacophilanthropy are no substitute for corporate acceptance of routine State restrictions on their profit in the interests of community welfare (Davies 2004). If GIPP continues to undermine the benefits of enhanced global trade by pursuing hegemonic objectives in such negotiations, it may eventually be forced to accept preeminent responsibility (financially and morally) for the global public health crisis it is creating.

What is urgently needed now are coordinated research projects in many nations accurately documenting the health and related regulatory impacts of trade-enforced increased intellectual property rights over innovative pharmaceuticals. Perhaps what is also necessary is for nations with pharmaceutical reference pricing systems, to group together to share research data and regulatory strategies that may foster and advance universal access to affordable, essential medicines as part of a basic commitment to bioethics and international public health. Perhaps they might even begin to include provisions facilitating such collaboration in their own bilateral trade deals, so that any subsequent renegotiation of TRIPS hears the many strong social justice and public health claims in this area.

NOTES

1. *Medicare Prescription Drug Improvement and Modernization Act* 2003 (US) para 101 (1860D-11(i) (codified as amended at USC para 1395w-111(j), and also paras 1101 and 1123 and conference agreement. Available at: <http://thomas.loc.gov/cgibin/cpquery/?&db_id=cp108&r_n= hr391.108 &sel=TOC_2588886&> (Last accessed: 11 February 2005).
2. (1980) 448 US 176.
3. 733 F 2d 858, 221 USPQ 937 (Fed Cir 1984).
4. World Trade Organization, Ministerial Declaration, Doha (WT/MIN (01)/DEC/1) US. Available at: http://www.wto.org/english/thewto_e/minist_e/min01_e/min01_e.htm (Last accessed: 7 April 2005).
5. World Trade Organization, Declaration on the TRIPS Agreement and Public Health (WT/MIN (01)/DEC/2) Available at: <http://www.wto.org/english/thewto_e/minist_e/min01_e/min01_e.htm> (Last accessed: 7 April 2005).
6. 307 F 3d 1351 (Fed Cir 2002).
7. *Madey v. Duke University*, 307 F 3d 1351 (Fed Cir 2002).
8. Petition for a Writ of Certiorari at 14, *Madey v. Duke University*, 307 F 3d 1351 (Fed Cir 2002) (No. 02-1007).
9. *Medicare Prescription Drug Improvement and Modernization Act* 2003 (US) New 42 USC 1860D-11 (i) as added by section 101 of HR1.

10. *Medicare Prescription Drug Improvement and Modernization Act* 2003 New 21 USC 355 (j) (5) as added by section 1101(a)(2)(D) of HR1.
11. *Medicare Prescription Drug Improvement and Modernization Act* 2003 New 21 USC 804(1) (1) (a) as added by section 1121 of HR1.
12. *Medicare Prescription Drug Improvement and Modernization Act* 2003 21 USC 108-173 as added by section 1123 of HR1.
13. *Medicare Prescription Drug Improvement and Modernization Act* 2003 21 USC conference agreement House Report 108-391 Title XI-Access to Pharmaceuticals. Available at: <http://thomas.loc.gov/cgi-bin/cpquery/?&db_id=cp108&r_n=hr391.108&sel=TOC_2588886&> (Last accessed: 11 February 2005); *Trade Act* 2002 (US), 107-210 §2102 (b) (8) (D).
14. *Trade Act* of 2002, 107-210, 2102(b)(8)(D).

REFERENCES

Abbott, F.M. 1998. The enduring enigma of TRIPS: A challenge for the world economic system. *Journal of International Economic Law* 1: 497-512.
———. 2002. The TRIPS-legality of measures taken to address public health crises: Responding to USTR-State-Industry positions that undermine the WTO. In: *The political economy of international trade law,* eds. D.M. Kennedy, and J.D. Southwick, 311-42. Cambridge: Cambridge University Press.
Attaran, A. 2004. How do patents and economic policies affect access to essential medicines in developing countries? *Health Affairs* 23(3): 155-66.
Bate, R. 2004. What patent problem? *National Review Online.* Available at: <http://www.nationalreview.com/comment/bate200405171342.asp> (Last accessed: 18 February 2005).
Beaumont, L. 2004. PBS advisor lobbying for US "a disgrace." *Sunday Age* 25 January.
Bekelman, J.Y., and C. Gross. 2003. Scope and impact of financial conflicts of interest in biomedical research: A systematic review. *JAMA* 289: 545-65.
Berlinguer, G. 2004. Bioethics, health and inequality. *Lancet* 364: 1086-91.
Chaudhuri, S.K., P.K. Goldberg, and J. Panle. 2003. *The effects of extending intellectual property protection to developing countries: A case study of the Indian pharmaceutical market.* NBER Working Papers 10159. *National Bureau of Economic Research.* Cambridge MA: National Bureau of Economic Research.
Cockburn, I., and R. Henderson. 1997. Public-private interaction and the productivity of pharmaceutical research. NBER Working Papers 6018. *National Bureau of Economic Research.* Cambridge MA: National Bureau of Economic Research.
Coffield, S. 1981. Using Section 301 of the Trade Act of 1974 as a response to foreign government trade actions: when, why and how. *North Carolina Journal of International Law and Technology* 6: 381-405.
Comanor, W.S. 1966. The drug industry and medical research: The economics of the Kefauver Committee investigations. *Journal of Business* 39: 12-18.
———. 1986. The political economy of the pharmaceutical industry. *Journal of Economic Literature* 24(3): 1178-217.
CP-Tech (Consumer Project on Technology). Big PhRMAs favorite academics and opinion makers. Available at: <http://www.cptech.org/ip/health/pharmadefenders.html> (Last accessed: 18 February 2005).
Corea, C.M. 2001. *Some assumptions on patent law and pharmaceutical R&D. Occasional Paper 6.* Geneva: Quaker United Nations Office.
Davies, G. 2004. *Economia: New economic systems to empower people and support the living world.* Sydney: ABC Books.
Deardorff, A. 1992. Welfare effects of global patent protection. *Economica* 59: 35-51.
DiMasi, J.A., R.W. Hansen, and H.G. Grabowski. 2003. The price of innovation: new estimates of drug development costs. *Journal of Health Economics* 22: 151-85.

Drahos, P., and D. Henry. 2004. The Free Trade Agreement between Australia and the United States undermines Australian public health and protects US interests in pharmaceuticals. *British Medical Journal.* 328: 1271-2.

Drahos, P., and J. Braithwaite. 2002. *Information feudalism: Who owns the knowledge economy?* New York: The New Press.

Drahos, P., T. Faunce, D. Henry, and M. Goddard. 2004. The FTA and the PBS. *Submission to the Senate Select Committee on the US-Australia Free Trade Agreement.* Canberra: Australian Senate.

Eisenberg, R.S. 1989. Patents and the progress of science: Exclusive rights and experimental use. *University of Chicago Law Review* 56: 1017-86.

Enyart, J. 1990. A GATT intellectual property code. *Les Nouvelles* 25: 53-6.

Faunce, T.A. 2005. *Pilgrims in medicine: Conscience, legalism and human rights.* Leyden: Martinus Nijhoff.

Fontanarosa, P.B., R. Drummond, and C.D. De Angelis. 2002. Postmarketing surveillance: Lack of vigilance, lack of trust. *JAMA* 292: 2647-50.

Gallini, N.T. 1992. Patent policy and costly imitation. *Randal Journal of Economics* 23: 52-63.

———. 2002. The economics of patents: Lessons from recent US patent reform. *Journal of Economic Perspectives* 16(2): 131-54.

Gathi, J.T. 2002. The legal status of the Doha Declaration on TRIPS and public health under the Vienna Convention on the Law of Treaties. *Harvard Journal of Law and Technology* 15(2): 291-317.

Global Alliance for TB Drug Development. 2001. *The Economics of TB Drug Development.* Available at: <http://www.tballiance.org/> (Last accessed: 15 April 2005).

Grabowski, H.G., and J.M. Vernon. 1983. *The regulation of pharmaceuticals.* Washington: American Enterprise Institute.

Grady, M.F., and J.I. Alexander. 1992. Patent law and rent dissipation. *Virginia Law Review* 78: 305-50.

Hettinger, E.C. 1989. Justifying intellectual property. *Philosophy and Public Affairs* 18(1): 31-52.

Irwin, D. 2005. *Free trade under fire.* Princeton: Princeton University Press.

Jaffe, A.B. 2000. The US patent system in transition: Policy innovation and the innovation process. *Research Policy* 29(4-5): 531-7.

Kastriner, L. 1991. The revival of confidence in the patent system. *Journal of the Patent and Trademark Office Society* 73(1): 5-23.

La Croix, S., and A. Kawaura. 1996. Product patent reform and its impact on Korea's pharmaceutical industry. *International Economics Journal* 10(1): 109-20.

Lanjouw, J.O. 1997. *The introduction of product patents in India: Heartless exploitation of the poor and suffering?* Growth Center Discussion Paper, Yale University and NBER Working Paper 6366. National Bureau of Economic Research. Cambridge MA: National Bureau of Economic Research.

Leary, V.A. 1994. The right to health in international human rights law. *Health and Human Rights* 1(1): 24-32.

Lerner, J. 2001. *150 years of patent protection,* Harvard MA, Harvard Business School Working Paper.

Lever, J. 1982. The new Court of Appeals for the Federal Circuit (Part I). *Journal of the Patent Office Society* 64(3): 178-208.

Médecins Sans Frontières, *Campaign for access to essential medicines.* Available at: <http://www.accessmed-msf.org> (Last accessed: 19 February 2005).

Metherell, M. 2002. Minchin's chief aide chosen as drug industry's cheerleader. *Sydney Morning Herald.* 17 February.

Mossinghoff, G. 1991. For better international protection. *Les Nouvelles* 26: 75-9.

Nogues, J. 1990. *Patents and pharmaceutical drugs; Understanding the pressures on developing countries.* Washington: World Bank Country Economics Department.

Nordhaus, W.D. 1969. *Invention, growth and welfare.* Cambridge, MA: MIT Press.

O'Donoghue, T., S. Scotchmer and J. Thisse. 1998. Patent breadth, patent life and the pace of technological progress. *Journal of Economics and Management* 7(1): 1-32.

Oxfam International. 2004. *Undermining access to medicines: Comparison of five US FTAs.* Oxfam Briefing Note. Available at: <http://www.oxfam.org.uk/what_we_do/issues/health/undermining_access_ftas.htm> (Last accessed: 15 April 2005).

Pazderka, B. 1999. Patent protection and pharmaceutical research and development spending in Canada. *Canadian Public Policy* 23(1): 29-64.

Peltzman, S. 1973. An evaluation of consumer protection legislation: The 1962 drug amendments. *Journal of Political Economy* 81(5): 1049-91.

PhRMA, Website. 2005. <http://www.phrma.org/issues/intprop/> (Last accessed: 18 February 2005).

Public Citizen. 2004. *The Medicare drug war: An army of nearly 1,000 lobbyists pushes a medicare law that puts drug company and HMO profits ahead of patients and taxpayers.* Washington DC, Congress Watch.

Ruger, J.P. 2004. Social determinants of health. *Lancet* 364: 1092-7.

Ryan, M. 1998. *Knowledge Diplomacy: Global Competition and the Politics of Intellectual Property* Washington DC: The Brookings Institution.

Sakakihara, M., and L. Dransteeter. 1999. *Do stronger patents induce more innovation. Evidence from 1988 Japanese patent law reforms.* NBER Working paper. Cambridge MA: National Bureau of Economic Research.

Scherer, F.M. 1984. *Innovation and growth: Schumpeterian perspectives.* Cambridge: MIT Press.

Scotchmer, S. 1991. Standing on the shoulders of giants: Cumulative research and the patent law. *Journal of Economic Perspectives* 5: 29-41.

Sell, S.K. 1998. *Power and ideas: The North-South politics of intellectual property and antitrust.* Albany, State University of New York Press.

———. 2003. *Private power, public law. The globalisation of intellectual property rights.* Cambridge: Cambridge University Press.

Shiner, J. 2004. Evidence to Committee on Finance Subcommittees on Health Care and International Trade. Washington: United States Senate.

Singer, P., and T. Gregg. 2004. *How ethical Is Australia. An examination of Australia's record as a global citizen.* Melbourne: Australian Collaboration and Black Inc.

Temin, P. 1980. *Taking your medicine: Drug regulation in the United States.* Cambridge: Harvard University Press.

Toebes, B. 1999. *The right to health as a human right in international law.* Antwerpen: Groningen.

US Department of Commerce, International Trade Administration. 2004. *Pharmaceutical price controls in OECD countries. Implications for US consumers, pricing, research and development and innovation.* Washington: US Department of Commerce.

USTR 1999. Special 301 review of South Africa policies on intellectual property rights. Available at: <www.uwstr.gov/releases/1999/04/99-41.html> (Last accessed: 21 February 2005).

———. 2004. Special 301 Report Watch List. Available at: <http://www.ustr.gov/Document_Library_ Publications/2004/2004_Special_301/200> (Last accessed: 7 September 2004).

Wehrwein, P. 2002. *Pharmacophilanthropy* Harvard School of Public Health. Available at: <http://www.hsph.harvard.edu/review/summer_pharmaco.shtml> (Last accessed: 7 April 2005).

WHO. 2000. Medicines Strategy: Framework for Action in Essential Drugs and Medicines Policy 2000-2003. Geneva: World Health Organization (WHO/EDM/2000.1).

Wiggins, S.N. 1983. The impact of regulation of pharmaceutical research expenditures: A dynamic approach. *Economic Inquiry* 21(1): 115-28.

CHAPTER EIGHT

PATRICIA PEPPIN

DIRECTING CONSUMPTION

Direct-to-consumer advertising and global public health

Produced by and for multinational corporations in order to promote products worth billions of dollars worldwide, direct-to-consumer advertising has the potential to produce adverse effects of a significant nature in the public health system. The promotion of pharmaceutical products through direct-to-consumer advertising has an impact on the doctor-patient relationship, health care expenditures and the health of patients. David Fidler (2002, 151) has noted that the processes of globalization, "including trade, investment, capital movements, travel, and technological advances – render borders more permeable and increase the volume, nature, scope, and complexity of problems governments face within their territorial boundaries." A globalized theory of health law, he has argued, takes us beyond national boundaries to examine national law, international law, and global governance, and at the third level – global governance – requires that we examine the role of non-state actors, including multinational corporations. For example, he has cited both the work of Médecins Sans Frontières in securing access to essential pharmaceutical products and the World Medical Association's development of ethics standards in the *Declaration of Helsinki*, which applies to pharmaceutical company research on human subjects (id., 157-8). The activities of such non-governmental organizations and multinationals restrain states' abilities to act in the public health arena and need to be taken into account in development of a globalized theory of public health law (id., 158).

The multinational drug companies operate in a global environment subject to national regulatory and litigation mechanisms. Effective multinational mechanisms are, however, largely non-existent. Direct-to-consumer advertising (DTCA) has become a controversial public policy issue in Western countries. Its advocates argue that DTCA educates patients and enhances patient empowerment, while improving health care. Its critics question its benefits and point to the limited information provided, the medicalization of health, the increasing amounts of health care dollars

B. Bennett & G.F. Tomossy (eds.), Globalization and Health: Challenges for Health Law and Bioethics, 109-128.
© 2006 Springer. Printed in the Netherlands.

going to fund prescriptions, and an altered doctor-patient relationship. The United States General Accounting Office Report summarizes the issue in this way:

> Supporters of DTC advertising maintain that it educates consumers about medical conditions and care options and that the increased use of prescription drugs that DTC advertising encourages has improved the public's health. Critics of DTC advertising contend that it is sometimes misleading, leads consumers to seek prescription drugs when other treatments may be more appropriate, and causes some patients to ask their physician to prescribe new drugs that are more expensive but may not be more effective than older drugs. Critics also argue that pharmaceutical companies spend too much money on drug promotion rather than on research and development initiatives.
>
> (GAO 2002, 1)

Because of its impact on health care spending and the doctor-patient relationship, such direct advertising raises questions about the commitment of governments to public health and brings into sharper focus the nature of the relationship of governments to the drug industry. In this chapter, I will begin by examining the regulatory context, particularly focusing on the United States and Canada, which have different legal and health care systems. Following this I will analyze its public health consequences through the commodification of the doctor-patient relationship, the construction of disease and the promotion of the medical model. Such promotion alters the doctor-patient relationship by transforming patients into consumers who are pre-sold on the quick cure of drugs in general and of advertised products in particular. Direct-to-consumer advertising produces significant public health consequences by increasing the consumption of pharmaceutical products, and this will be the subject of the last section.

THE REGULATORY CONTEXT

Only New Zealand (Toop et al. 2003, 1) and the United States, among the thirty member countries of the OECD (Organization for Economic Cooperation and Development), permit direct-to-consumer advertising of prescription drugs. New Zealand permits such advertising in the absence of any provisions in their governing legislation, the *Medicines Act* 1981 (ibid.). Because New Zealand provides for public payment of drug costs through a prescription benefit scheme under their national health service, the economic implications of DTCA differ significantly in that country.

In the United States, with the "clarification" of the guidelines governing direct-to-consumer advertising in 1997, industry received the signal to vastly increase the amount of direct-to-consumer advertising. By 2001, DTCA spending amounted to US$2.7 billion, an increase of 145 percent in four years (IMS Health 2002; GAO 2002, 3, 9-10). This increase in consumer-directed advertising, according to a study by Pinto and colleagues, was a response to several factors preventing market expansion, including the unchanging prescribing patterns of doctors and the limitations on prescribing imposed by managed care managers, in combination with patients' expectations of a role in decision making and increasing pressures from drug discounters and generic manufacturers (Pinto, Pinto and Barer 1998; Mintzes 2001, 7).

The Food and Drug Administration (FDA) had issued a draft guidance on broadcast advertising clarifying the regulations under the Federal *Food, Drug and Cosmetic Act*,[1] and how they applied to broadcast ads. Print ads are required to disclose all risks listed in the approved product labeling (the "brief summary"). In addition, the "fair balance doctrine" requires that a balanced view of risks and benefits be presented in the ad as a whole (Palumbo and Mullins 2002, 428). The draft guidance permitted the industry to leave out the "brief summary" of labeling information; instead, broadcast ads may meet the regulatory requirement by disclosing the major risks of the product in the audio or audio and visual parts of the ad and, as an alternative to disclosing the brief summary, by making "adequate provision" to direct patients to alternative sources of information where complete information may be obtained (FDA 1997; 1999 and updates). Certain assumptions made by the FDA about these ads are specified in the guidance, as are the means by which companies may meet the adequate provision requirement, such as toll free numbers, asking your doctor, pharmacist or other health care provider, an Internet website, and available print ads (GAO 2002, 6-8; Palumbo and Mullins 2002, 431). This change made broadcast ads affordable, since the "brief summary" had been expensive, and more attractive, since the message had been stylistically obtrusive in the previous format. Broadcast ads had contained so much information about risks and contraindications that it must have been clear to patients that drugs have risks – not a desirable marketing message.

Australia, New Zealand, the European Union and Canada have all been engaged in reviews of the issue of DTCA, considering whether to permit DTCA by legislation or, in New Zealand's case, whether to limit it by regulating or banning it (Mintzes 2001, 12-22). Australia and the European Union have rejected proposals to change to a more permissive system. The European Parliament recently voted against a proposal to permit advertising directly to consumers for drugs for HIV/AIDS, asthma and diabetes, by 494 to 42, a 12:1 ratio against (Health Action International 2002; Toop et al. 2003, 39-40). New Zealand and Australia signed a treaty in late 2003 to create a single agency to regulate drugs, devices and complementary medicines for both countries (Burton 2004, 1036). On the basis of a submission by the health minister to ban DTCA, the New Zealand cabinet decided that the health minister should discuss harmonizing their standards with Australia, which prohibits DTCA but permits materials promoting disease awareness without brand names (ibid.).

The Canadian government has begun a consultation on a draft *Canada Health Protection Act* that includes four options in this area, including legalization of DTCA. The Canadian legislation, the *Food and Drugs Act*, prohibits fraud and misrepresentation. While advertising may be directed only to prescribing professionals for Schedule A and F drugs (prescription medicines), Health Canada has permitted a certain amount of DTCA through a loose interpretation of a 1978 exception permitting direct advertising of name, price and quantity, a Regulation intended to permit pharmacists to advertise competitive prices, and through interpretation of the meaning of the terms "education" and "advertising" (TPP 2002). The limited scope of this exception is quite clearly stated: "Where a person advertises to the general public a Schedule F Drug, the person shall not make any representation other than with respect to the brand name, proper name, common name, price and quantity of the drug."[2] The period of expansion of American DTCA saw the Canadian government permit direct advertising of images and text constituting representations other than name, price and quantity, and reminder and help-seeking ads in which a disease or condition is discussed but no specific prescription drug. These expansions, without any form of public debate, have clearly gone beyond the scope intended for the provisions. As well, advertisers have pushed the boundaries of the word "advertisement," claiming the educational value of their promotional activities. Canadians have been exposed to a significant amount of direct advertising because of the penetration of the US broadcast media and the significant cross-border trade in periodicals, as well as through domestically produced ads, and it appears that the Canadian government has been "testing the waters" before making a political move. The Internet has greatly increased access to sites for information, promotion and purchase of prescription drugs, to those with hardware, software and Internet connections.

The current system in Canada relies on preclearance of physician-directed prescription drug advertising by the Pharmaceutical Advertising Advisory Board (PAAB),[3] an arm's-length body made up of industry representatives, health practitioner organizations, journal editors and a consumer organization. Health Canada is not a member but sits ex-officio. Companies in the "research-based" pharmaceutical company organization have agreed to adhere to the PAAB standards and have also created a set of standards of their own to apply to their members. The *Competition Act* provides an additional forum to address intercompany disputes of an anti-competitive nature. Such self-regulation is attractive to pharmaceutical companies as a means of controlling their competitors. The PAAB carries out responsibility for applying federal regulatory standards and consults with Health Canada officials in doing so. Infractions serious enough to constitute criminal offences could be prosecuted by the federal government under the *Food and Drugs Act*, although voluntary compliance is the means of regulation preferred by the federal government – an approach consistent with government's strong support for industrial development.

New Zealand also has a preclearance system operating through the self-regulating Therapeutic Advertising Pre-vetting System (TAPS) managed through the Association of New Zealand Advertisers but increasingly incorporating delegated authorities to companies to approve changes to their own ads (Toogoolawa 2002, 8; Toop et al. 2003, 28-30; Mintzes 2001, 14). The process has the capacity to react quickly if an ad is considered to endanger public health or seriously mislead as the participating publishers can prevent further publication, although not all publishers are part of the system (Toogoolawa 2002, 8-9). Breaches are investigated only in response to complaints (Toop et al. 2003, v, 31). A report on DTCA prepared by family medicine physicians from all four New Zealand Departments of Family Medicine identifies the weaknesses of the New Zealand system, including a lack of independence of the pre-vetting and complaint response systems, which consist of advertising and pharmaceutical industry representatives; lack of any independent technical review; lack of detailed criteria for content and presentation; lack of regular compliance monitoring by a central independent agency; lack of an effective deterrent; and the need for any complainant to possess knowledge about the risks and benefits of new medications to be able to carry out a comparison with the advertisement (id., 30). The Toogoolawa review described the system as a strong self-regulatory system, based on the strong leadership of the advertising and media industries (Toogoolawa 2002, 53).

The Australian system of therapeutic goods advertising is also co-regulated, operating under the legislative authority primarily under the *Therapeutic Goods Act* 1989 and with industry participating both through self-regulation and through the exercise of delegated legislative authority (id., 37-52). The system specifies certain higher-risk therapeutic products that may be advertised only to medical practitioners, and most of these must be approved prior to publication. Some of the remaining products, excluding prescription medications and those sold only through pharmacies, may be advertised to consumers in certain media and these are either subject to preapproval or publishable with only name, picture and/or price and/or point of sale and without therapeutic claims (id., 37-39). Certain kinds of advertising to consumers require preapproval by industry associations acting under delegated authority while other advertising, including physician-directed ads, is covered by industry self-regulation through the industry code of conduct (Toogoolawa 2002, 9, 39, 46; Mintzes 2001, 19). The Toogoolawa Consulting Pty Ltd Report on the Australian and New Zealand systems review, carried out in the context of the proposed Trans-Tasman Agency, states that the Australian system, and the preapproval and complaints systems in particular, has been seen by most as "unduly complex, open to inconsistency and often too slow to protect the public interest when it comes to public health and safety" (Toogoolawa 2002, 9). DTCA was explicitly excluded from the terms of reference (id., 8).

In the United States, the FDA regulates prescription drug advertising under the Federal *Food, Drug and Cosmetic Act*.[4] This legislation prohibits misbranding, which consists of making false and misleading claims, failing to present risk-benefit information in a fair manner, and otherwise failing to meet regulatory standards (GAO 2002, 2). The FDA reviews advertising on the basis of materials filed by manufacturers. Companies are required by regulation to send their advertisements to the FDA when they are first distributed to the public, and the FDA assessors examine these submitted ads for conformity with federal regulations (id., 2, citing 21 CFR 314.81(b)(3)(i)). Unlike the Canadian and New Zealand systems, this system operates through review of the ads by the government agency and this review commences after the ad has been distributed rather than prior. Because of the nature of their regulatory authority, ads are generally not approved in advance, but companies may ask for advisory comments prior to an ad campaign and the FDA has asked that companies voluntarily submit DTCA prior to distribution (Palumbo and Mullins 2002, 429; GAO 2002, 17). The Division of Drug Marketing, Advertising, and Communications within the Center for Drug Evaluation and Research (CDER) reviews these promotional materials. While specific figures for direct-to-consumer ads reviewed were not available, the FDA reported that they had received approximately 34,000 pieces of promotional material directed to physicians and consumers in 2001 (GAO 2002, 17, n27). The Division sends regulatory letters to those it considers to have violated federal standards, amounting to about 5 percent of broadcast ads reviewed from 1999 to 2001, and eighty-eight such regulatory letters for direct-to-consumer ads in the five year period from August 1997 to August 2002, almost all considered to have been for the less serious level of violation (id., 4, 18). These letters requested that the company stop distributing the misleading DTC ad and led to the companies doing so in every case (id., 21). Other remedies available are legal action to seize drugs, a request to the court to stop the ads and impose a corrective ad campaign, or imposition of criminal penalties (ibid.). The relatively weak penalties imposed in this system, like the other systems, is an indication of the government's own position.

The US General Accounting Office Report, submitted to Congress in October 2002, concluded "While generally effective at halting the dissemination of advertisements it reviews and identifies as misleading, FDA's oversight of DTC advertising has limitations" (id., 4). Not all companies comply with this regulation. Some companies fail to submit their ads or fail to submit them in accordance with the time lines (and it is difficult to find out which ones haven't been submitted), while some companies have repeatedly distributed new misleading advertisements for a drug after receiving a letter about the same drug. The number of staff engaged in this exercise is small – five people filling seven possible positions to review DTCA (id.; Toop et al. 2003, 34). The GAO expressed concern about a recent change in FDA policy applying to the sending of regulatory letters to companies identified as being in violation: "FDA's oversight has been adversely affected by a January 2002 change in its procedures for reviewing draft regulatory letters that was directed by the Department of Health and Human Services (HHS)" (GAO 2002, 4). This change in the review process requiring review by the Office of the Chief Counsel for "legal sufficiency and consistency with agency policy" (id., 22) before sending regulatory letters to companies has significantly increased the time between FDA identifying an ad violation and the request to remove it. In response to the draft Report, Health and Human Services replied that the change was adopted in order to ensure that before regulatory letters were sent to companies, the letters rested "on a sound legal foundation, are credible, and will promote compliance" (id., App III, 22, 24). In the Report, the GAO indicated that HHS had agreed to speed up the process (id., 24). This slowdown in the drug review process was seen as significant because it increased the time during which consumers would be exposed to misleading advertising. The ad campaign might be over before the regulatory letter arrived.

These systems are clearly inadequate to meet the deluge of problems created by direct-to-consumer ads.

THE IMPACT OF DIRECT-TO-CONSUMER ADVERTISING
ON PUBLIC HEALTH

The idea that ads promote consumption seems obvious. As viewers of ads for consumer goods such as soft drinks, running shoes and cars, we are aware of being targeted by advertisers. Now that drug company spokespersons claim that ads educate, we need to examine the role played by drug advertising.

The mechanisms by which advertising promotes product consumption have been the subject of study by semiotic theorists, from Roland Barthes to modern analysts Judith Williamson and Robert Goldman (Barthes 1972; Williamson 1978; Goldman 1992). A sign is defined as a thing (a material object, person, drawing) plus its meaning. The material object is the signifier, which is combined with what is signified or connoted to form the sign. Signifier plus signified=sign. "Good morning!" sings the happy man as he leaps into the air above the white picket fence depicted on our television sets. The sign of the man includes the signification of happiness created by Viagra, while the white picket fence provides a further sign whose meaning is the normalcy in middle class America of erectile dysfunction and adoption of the drug remedy. Advertisers place in the ads signs whose meanings will be interpreted by the viewers and transferred over to the product to be consumed.

If semiotic analysis is applied to ads directed to doctors, we see that the doctor is the viewer who interprets the ad's signs, providing the source of meaning for the ad, the referent system. As expressed by Goldman and Papson (1996, 9):

> Producing marketable commodity signs depends on how effectively advertisers are able to colonize and appropriate referent systems...Any referent system can be tapped, but remember that advertisers appropriate referent systems for the purpose of generating sign value, so they dwell on referent systems that they calculate might have value to their target audience...The referent systems that can pay off most handsomely when properly appropriated involve lifestyle and subcultures. In recent years advertising has appropriated nostalgia, hip-hop music, grunge, and feminist sensibilities.

In advertising directed to physicians, advertisers attempt to discern what doctors will see in the ads, often based on the political culture of medicine and broader social values as well as nature and science (Peppin and Carty 2001a). As the population of doctors becomes more pluralistic and representative of the broader social groups in the US and Canada, and as traditional views of medicine change, advertisers must find discerning the group norms more difficult.

Advertisers structure the transfer of meanings from sign to commodity, creating commodity signs (Williamson 1978). Children wonder what's going on when they see the "Good morning!" ad, but adults have no problem transferring the sexual happiness of the singing man to the Viagra and then over to the consumer, who will be happy too if he uses the product.

In research on physician-directed ads carried out with Elaine Carty, we have examined ways that ads convey messages about the disease itself, structuring perceptions that particular conditions are diseases, such as menopause (Peppin and Carty 2001b) and depression (Peppin and Carty 2001a; 2003). The ads portray disease in a simplified and acontextual way that ignores the social determinants of health. They use social stereotypes – of gender and disability, for example – to portray patients, in this way perpetuating biases against particular groups and promoting disparate perceptions of patients with the disease. Over the fifteen years of ads for the hormone replacement therapy Premarin, we observed the changes in the social stereotype of the patient from pathetic vulnerable woman to uppity female seeking outside information to jogging baby boomer concerned with bones and hearts to a cracked pelvis portrayed in an x-ray. None of these portrayals suggested an autonomous woman. They promote intervention in general, drug intervention in particular. All these effects are likely to continue in direct-to-consumer advertising. As ads use stereotypes to transfer meanings, they also structure perceptions of the patients with the disease who need the drug. In creating such a diagnostic image, we have argued, the images of the patient for the drug have been over-inclusive of groups on the one hand and under-inclusive of groups on the other, with likely effects on diagnosis and prescription (Peppin and Carty 2001a, 573-4). Such a trend is likely to continue with DTCA. Self-diagnosing that results from viewing direct-to-consumer ads is likely to replicate the population displayed in the ad and to perpetuate the stereotypes.

Once advertising of prescription drugs is directed to patient/consumers they take on a new role as participants in the ad process for prescription drugs and interpret and contribute meaning to the images provided by the industry. Patients have become experienced participants in the ad process through their engagement with consumer culture. As Goldman and Papson (1996, 3) have pointed out, this process begins at a very young age.

> Product standardization makes it imperative that products attach themselves to signs that carry an additional element of value. Nike captured a larger market share of the sneaker industry than Reebok did between 1986 and 1993 because Nike effectively harnessed the power of Michael Jordan's image while Reebok failed to counter with a superior or even an equal stream of imagery. In this kind of industry, everything depends on having a potent, differentiated image.

With the creation of "me-too drugs," replicas of existing products, creation of this additional value becomes essential to drug branding.

"Ask Your Doctor" is a common instruction in US broadcast advertising. This message itself has layers of meaning. The instruction to patients to ask questions of their doctors fits within our legal and ethical understanding of the doctor-patient relationship, which relies on communication by the doctor and increasing comprehension by the patient so that the patient may make an autonomous decision about the proposed treatment. This relationship of increased equality and communication between the doctor and patient has been nurtured by the courts in the US and Canada, through judgments requiring disclosure of information about material risks and alternatives and assessment of the patient's comprehension. When the pharmaceutical manufacturer is added to the relationship, the manufacturer is at the top of the hierarchy of information, with each lower rung relying on the rung above it for information. As the Supreme Court of Canada in *Hollis v. Dow Corning Corp.*[5] and the Ontario Court of Appeal in *Buchan v. Ortho Pharmaceuticals*

(Canada) Ltd.[6] have observed, the relationship is one of inequality based on the vastly superior informational advantage possessed by the pharmaceutical manufacturer and the limited resources of patients and even of doctors. The doctor acts as a learned intermediary between the manufacturer, from whom the doctor receives product information, and the patient, to whom the doctor discloses information.

Directing patients to their doctors enables an American manufacturer to dispense with complete discussion of risks in the broadcast ad copy. This phrase constitutes one acceptable means of avoiding insertion of the brief summary into broadcast ads by complying with the "adequate provision" standard for broadcast advertising under the regulatory clarification of 1997. Manufacturers may place less risk information in their ads and in this way change the perceived balance between benefits and risks. The instruction serves in this way as an instrument in the marketing of the product.

The meaning of the statement also includes asking your doctor for this product. This instruction has been taken to heart by patients. The GAO Report stated that, "The percentage of patients asking their physicians about a prescription for a specific drug was consistent across studies, about 30 to 35 percent of those who remembered seeing a DTC advertisement" (GAO 2002, 16, App II). Barbara Mintzes' extensive literature review indicates that about a quarter of patients surveyed spoke to their doctor about an advertised drug, about 5 to 7 percent of all patients asked for the drug and about four-fifths of these received the prescription (Mintzes 2001, Table 6, 36; studies were carried out by Prevention (1998; 1999; 2000), Time (1998) and National Consumer League (1998)).

The GAO estimated that 8.5 million consumers would have received a prescription, based on a 5 percent figure and a total of 170 million adults visiting a doctor in a year (GAO 2002, 16).

Patients not only ask their doctors but also are seen by doctors to be putting pressure on them and expecting results. Doctors feel pressured to provide the particular drugs requested, as surveys cited by the New Zealand family physicians indicate (Toop et al. 2003, 19-21). Doctors try to keep their patients happy and respond to patient requests to try a new treatment and also feel concerns about losing patients (id., 19, citing Avorn, Chen and Hartley 1982 and Cockburn and Pit 1997). A study by Bell, Wilkes and Kravitz (1999) of Sacramento patients' expected responses to physician denial of an ad-induced drug request indicated that patients would respond by attempting to persuade the doctor or go prescription shopping in a quarter of the cases. Negative reactions correlated with positive attitudes to DTCA and undue confidence in government regulation (ibid.). The New Zealand doctors conclude that, "These commercial pressures make it increasingly difficult for the modern GP to balance both patient-centred and evidence-based medicine" (Toop et al. 2003, 20). The comparative study by Mintzes and colleagues of patients and doctors in Sacramento and Vancouver, found that "more advertising leads to more requests for advertised medicines, and more prescriptions" (Mintzes et al. 2003, 405). A seventeen-times increase was observed and nine of ten received drugs (id., 411). Any conversation initiated by the patient is "highly likely to end with a prescription, often despite physician ambivalence about treatment choice" (id., 405). This study is particularly significant because it used a control group, which had not been exposed to DTCA to the same extent as the study group.

In this model of doctor-patient interaction, the patient has become the agent of demand. The patient drives the economic decision – and creates a psychological demand. Patients themselves become "pre-sold" consumers rather than patients discussing information about a proposed treatment. Even the language recognizes this shift – it's direct-to-consumer advertising. Similarly, the literature on this question contains numerous plays on the metaphor of "who is minding the store?" Asking your doctor in this sense means transforming the relationship from a doctor-patient model of caring advice and mutual decision making into one of consumer and gatekeeper to the drug supplies.

Efforts have been made by the industry to link their direct advertising efforts to the movement for greater autonomy and a participatory doctor-patient relationship. For example, the PhRMA website states that DTCA, in addition to promoting products, "helps educate patients," "encourages dialogue" between patients and doctors and is "being widely employed" as patients seek out information and move towards "a more patient-focused health care system." As an article in the *American College of Physicians – American College of Internal Medicine Observer* expressed it:

> Drug companies claim that advertising promotes consumer education and stimulates dialogue between physicians and patients – and some doctors agree. They say that patients are now more willing to discuss symptoms they've seen referred to in advertisements....
>
> But even more physicians say that TV commercials do not improve patient communication. Dr. Gutner from Massachusetts, for instance, said that only a minority of his patients – perhaps 20% – are really interested in discussing an advertised treatment beyond asking for a prescription. "The dialogue doesn't get to where I'd like to see it go," he said.
>
> Instead, doctors are fielding requests. "It's altered the dynamics within the [exam] room," said Joseph G. Weigel, FACP, the College's Governor for the Kentucky Chapter. "Patients almost feel that the physician's office is the drive-through window at McDonald's where they put their order in and you fill it."   (Maguire 1999, online 2-3)

It is understandable that physicians would find themselves uncomfortable in their new role. When patients ask for particular products, several steps in the process have been skipped and the physician may find it necessary to persuade a patient that a particular product is unnecessary or a poor alternative. Changing perceptions may be difficult and the doctor may feel reluctant to displease the patient, particularly in a fee-for-service system, or may not have the time to engage in the full process of undoing the message. These views were all expressed in the study carried out by the New Zealand family physicians. In addition, doctors need to recognize that their prescribing patterns are influenced by the pharmaceutical industry through means directed at them, and to take steps to reduce this reliance, as Breen has argued

(Breen 2004). Doctors may be cross-pressured by direct and mediated drug messages and feel a need for independent and reliable sources of information.

Margaret Gilhooley has noted the special pressures to prescribe one drug resulting from the "single-drug focus" of DTCA and has recommended that direct-to-consumer ads contain a prominent statement that consumers "need to 'Consult your doctor about the range of treatment choices that may be available, and their risks, benefits and costs.'" In her opinion, this disclosure would fit within the category of disclosure to prevent consumer deception, which constitutes an exception to the constitutional protection of free speech (Gilhooley 2003, 919).

Do patients recognize diseases and seek help? It seems fair to say that many individuals are unaware that they have real illnesses such as diabetes or cardio-vascular disease and asking one's doctor could lead to diagnoses of problems. There were some indications of this in the FDA study released in early 2003 but methodological limitations of this survey have been noted (Mitka 2003, 827-8; Mintzes et al. 2003). The New Zealand doctors stated that even if DTCA leads to more visits to doctors, the evidence does not indicate that disease is detected earlier or that patients are more likely to take the drugs (Toop et al. 2003, 16). This is clearly an area that requires carefully designed independent studies. Even if this is correct though, is DTCA the best way to achieve patient recognition of symptoms? More attention paid to preventive medicine seems likely to achieve similar results. As indicated above, industry identification of symptoms is highly contingent on market desirability of the "disease." Further, many requests for prescriptions are for lifestyle conditions or symptoms (Mintzes et al. 2003, 9).

Research provides indications of limitations on the information conveyed. Bell, Wilkes and Kravitz (2000) conducted a study of 329 consumer ads appearing in eighteen popular magazines from 1989 to 1998. Coders assessed whether the ad contained information about the specific aspects of the condition for which the drug was advertised and the treatment. The authors found that most ads provided information about the disease's symptoms but that few provided information about what had led to the condition, its prevalence, misconceptions about the condition, or about the drug's success rate, mechanism of action, treatment length, alternatives to the drug or behavioral changes that the patient could make to improve their health. They concluded that while informative ads were found, most provided minimal information and that the educational value of advertising could be improved.

The extensive literature analyses carried out by two groups, the University of British Columbia Centre for Health Policy and Research and the New Zealand Report group of family physicians from the four New Zealand Family Medicine departments, found that there is no evidence to indicate that patients are better informed as a result of DTCA. Evidence to the contrary exists in studies of physician-directed advertising, in the number of warning letters sent out by the FDA for regulatory violations in conveying risks and product information by DTCA advertisers and by the frequency of repeat offenders. In the period immediately following the guideline clarification by the FDA, a period of about a year and a half from August 1997 to early 1999, seventeen of thirty-three broadcast ads – just over half the broadcast ads – were found to have violated the US Federal *Food, Drug and Cosmetic Act* (Mintzes 2001, 65). For example, the FDA issued written notices to Bristol-Myers Squibb which used women athletes in ads promoting their drug Pravachol to reduce heart attack risk, although the drug had not been tested on women, and to Pharmacia & Upjohn for omitting from their ad for Caverjet, for erectile dysfunction, the information that the drug must be injected into the penis (Pear 1999; Working Group on Women and Health Protection 1999, 2). AstraZeneca received a regulatory letter from FDA in 2001 with respect to its breast cancer drug Nolvadex (tamoxifen citrate), for making misleading efficacy claims, minimizing risks and failing to comply with postmarketing reporting requirements (GAO 2002, 20, Table 4).

The New Zealand report concludes, in one of its ten key findings, that DTCA:

> does not focus upon promoting choice, but instead upon creating demand for specific medicines. This conflicts with the right of the patient to have easy access to high-quality, independent, comparative information on the risks and benefits of available pharmacological and non-pharmacological treatments. The information DTCA provides does not follow any accepted guidelines for health promotion or provision of consumer health information.   (Toop et al. 2003, vi).

The New Zealand family physicians argue that there is a need for independent consumer information and recommend that an independent medicine and health information service be established.

Sidney Wolfe, director of the Public Citizen Health Research Group, has said in an editorial in the *New England Journal of Medicine*: "The education of patients – or physicians – is too important to be left to the pharmaceutical industry" (Wolfe 2002, 525). He makes the important point that physicians who are misled by advertising – as research has shown they are – will act as "duped gatekeepers [who] may not adequately resist patients' exhortations to write a prescription." Wolfe called for the National Institutes of Health and the FDA to "replace tainted drug company 'education' with scientifically based, useful information that will stimulate better conversations between doctors and patients and lead to true empowerment" (ibid.).

Margaret Gilhooley has also observed that better data may be needed to assist physicians in evaluating drugs on the basis of their relative efficacy and noted that a research institute has been proposed as a means of achieving this (Gilhooley 2003, 921; Reinhardt 2001, 147). An independent source of medical information would be a vital resource in educating patients and promoting a strong participatory relationship between doctors and patients. Better information compiled by an independent source is essential for patients to have power over their own bodies. In advertising to doctors and to consumers, the same channel of information is used to distribute information about the product and to sell the product. The entwining of the commercial purpose with the education purpose creates a conflict, with the potential to undermine the disclosure legally required of the pharmaceutical industry under tort law. Careful vetting prior to broadcasting is required to ensure that individuals are protected from excessive or improper claims. In some cases such excesses would amount to fraud or misrepresentation, which could bring criminal charges or civil action. While commercial speech is constitutionally protected in both Canada and the United States, there are limits on such protections. As noted above, Margaret Gilhooley has recommended that the FDA consider the need for better information to prevent deception as the FDA reassesses DTCA (Gilhooley 2003, 921). The United States Supreme Court recently decided a case involving fraud, in which they determined that a fraudulent charitable solicitation, "like other forms of public deception" would not be constitutionally protected under the First Amendment.[7]

When risk information is removed from advertising, it becomes less apparent that products have risks. In studies by *Prevention* magazine, it was found that, "49-55 percent believe that DTCA makes drugs seem harmless." Even if the risk-benefit information were presented more fully and accurately in the ad visuals or TV voice-over, ads would not be the best vehicle to communicate this information since they are vehicles of consumption. Ads communicate through more complex means. Ads in popular magazines and on television sell through minimizing information and maximizing lifestyle content. Messages about lifestyle, patient populations, and social characteristics – including stereotyping of groups – all form part of medical advertising. As in other forms of consumer advertising, lifestyle equivalences are set up so that viewers can equate consumption with success. Is taking Prozac or Paxil now a lifestyle – or Viagra – or Diane-35?

Drug advertising contributes to the creation of diseases. It sells the message that particular diseases exist – such as hormone "deficiencies" of menopause requiring HRT or attention deficit disorder requiring Ritalin, or panic disorder requiring an SSRI, or premenstrual dysphoric disorder also requiring an SSRI. The medical model suits drug producers and promotion creates images of diseases that seemingly arise from and respond to biochemistry – "mind mechanics" making the brain amenable to antidepressants (Goldman and Montagne 1986).

In an editorial in the *British Medical Journal*'s special issue on medicalization, journalist Ray Moynihan and editor Richard Smith put the case for de-medicalization, noting how slippery the concept of disease is and suggesting that a de-medicalization agenda might include transferring power back to patients, encouraging self care and autonomy, more equitable worldwide access to basic health care, resisting categorizing life's problems as diseases, de-professionalizing primary care and deciding which complex services should be made available (Moynihan and Smith 2002, 860). In that issue, Moynihan, Heath and Henry (2002, 886) provide a critical analysis of disease-mongering, noting that, "The social construction of illness is being replaced by the corporate construction of disease." They give five examples: ordinary processes or ailments as medical problems: baldness; mild symptoms as portents of serious disease: irritable bowel syndrome; personal or social problems as medical ones: social phobia; risks conceptualized as diseases: osteoporosis; and disease prevalence estimates framed to maximize the size of a medical problem: erectile dysfunction (ibid.). The contingent nature of disease is perhaps most evident when we examine it cross-culturally. For example, depression is rampant in Western democracies while almost non-existent in other parts of the world. As David Healy (1997) has indicated, depression has been constructed as a disease largely through the efforts of the pharmaceutical industry, which also creates new categories such as social anxiety disorder and panic disorder to expand the diagnosis and the market for SSRIs (Medawar 1996 updated, paras 2.4, 2.5). Anthropologist Margaret Lock has written about the contrasting experiences of menopause of women in Japan who do not experience menopause as the same kind of event and have few symptoms of the type recognized in North America (Lock 1993).

Patients have been found to hold trusting and incorrect views about the extent of government regulation. In a survey of Sacramento County residents, Bell, Kravitz and Wilkes (1999) assessed "misplaced faith in advertising" by asking whether particular statements were true or false. Substantial numbers of respondents believed that DTC ads were assessed by the FDA prior to broadcast (50 percent) and further, that only the safest and most efficacious drugs would be allowed to be advertised (43 percent) (ibid.). These are not unreasonable expectations – but they are incorrect. The New Zealand study concluded that this study, along with other surveys finding that people believe that ads have been reviewed or endorsed by government, led them to believe that there is "a false belief in regulatory protection" and inadequate risk communication (Toop et al. 2003, 12, citing Slaughter 2001-02; Mintzes et al. 2003; NIHCM 2000). Excessive ill-founded reliance and inadequate risk disclosure is a dangerous combination, especially when allied to the techniques of persuasion that semiotic analysis exposes. Such beliefs may be based in naïve faith in government, but may also be nurtured by ads promoting drug miracles and the myths of scientific legitimacy.

HEALTH CARE COSTS RELATED TO DTCA

Advertising expenditure and purchases of drugs have escalated since the 1997 clarification of the US guidelines. The increase in retail spending on prescription drugs from 1999 to 2000 in the United States was US$20.8 billion. The fifty drugs most heavily advertised to consumers made up 48 percent of that amount and the other 9,850 drugs made up the other 52 percent of the US$20.8 billion (NIHCM 2000, 2, 4, 9 and Figures 2, 5). The top ten DTCA advertising budgets in 2000 were for arthritis (two drugs), allergy (three), ulcer/reflux, anxiety/depression, high cholesterol, impotence and obesity: Vioxx (arthritis), Prilosec (ulcer/reflux), Claritin (allergy), Paxil (antidepressant) and Zocor (high cholesterol) were the top five (id., 8). The top five each had an ad budget in excess of US$91 million. Advertising is concentrated on drugs for chronic conditions (NIHCM 2000, 12) – drugs that will be taken for a long time – and drugs for common conditions (GAO 2002, 12-13).

Drug promotion spending was US$19.1 billion in 2001, with direct-to-consumer advertising accounting for US$2.7 billion of that amount. DTCA increased by 145 percent from 1997 to 2001 (IMS Health 2002; GAO 3, 9-10). Increases were also observed in research and development spending but by only 59 percent. Even as direct advertising has become more extensive in its reach, it is significant that the industry still spends the majority of its advertising budget promoting its products to doctors. In 2001, the expenditure on journal advertising alone in the US was US$425 million (Rosenthal et al. 2002). Rosenthal and colleagues reported that even though DTC advertising of almost US$2.5 billion had grown disproportionately to other promotional efforts, it made up only 15 percent of drug promotion (Rosenthal et al. 2002).

It might be argued that such expenditure is justifiable based on the innovation of the pharmaceutical industry. Such an argument would not be persuasive though. A study by the independent NIHCM Foundation analyzed the drugs approved by the FDA in the period from 1989 to 2000 (NIHCM 2002). Using the FDA's categories, they found that, "Highly innovative drugs – medicines that contain new active ingredients and also provide significant clinical improvement – are rare," making up only 15 percent of new drug approvals during this period – 153 of 1035 new drug approvals (NDAs) (id., 3). During the latter half of this period, when priority reviews were granted to some drugs, only 13 percent of new drug approvals were priority new molecular entities (NMEs). In spite of this relatively low rate of innovation, the study noted that, "From 1995 to 2000, total retail prescription drug spending more than doubled from an estimated $64.7 billion to $132 billion" (id., 10). Two-thirds of this spending came from new (1995 onwards) products, two-thirds of increased spending for new products came from standard rated products rather than those given priority review status, and all the new drugs were "significantly higher in price" (id., 10-1). This lack of innovation has another effect. Major companies depend on innovation for future profits and the multinational industry needs drugs that will make big profits – "blockbuster drugs." If the pharmaceutical companies cannot count on innovative drugs to make big profits then they need to expand the markets for the products they already have (Medawar 2001). This means DTCA.

Arnold Relman and Marcia Angell, former editors-in-chief of the *New England Journal of Medicine*, have pointed to the high degree of public contribution and relatively small industry contribution to research (Relman and Angell 2002). Publicly funded research is supported by "government-granted monopolies – in the form of patents and FDA-approved exclusive marketing rights" (id., 28). They cited an unpublished National Institutes of Health document that analyzed the top five drugs in 1995 sales, a group that included Zantac and Prozac. All but one of the seventeen key research papers leading to these drug developments came from outside the industry and 85 percent of all relevant published research came from labs supported by public funds or foreign labs. Even taking into account the greater incentive for academics to publish, they concluded that publicly funded research rather than the pharmaceutical industry is the major source of innovation (id., 30).

Does direct-to-consumer advertising produce benefit? The evidence does not support this. As the New Zealand Report notes:

> Little evidence has been found that suggests a net benefit to the public health system. On the contrary, the evidence actually suggests net harm to public health and a serious risk to the financial sustainability of health systems where DTCA is permitted.
>
> (Toop et al. 2003, iii)

There is evidence of adverse effects. If the advertised product is a new drug, then the adverse drug reactions will remain partially unknown at the time they reach the market, simply because sample sizes during clinical trials are relatively small and perhaps unrepresentative of the population in which they will be used. The product can achieve significant market penetration, with many people exposed, before the risks become apparent. Existing medications are sufficient in most instances and the adverse effects that occur after a drug is introduced to the markets can be avoided by using existing medications (id., 22). If direct-to-consumer advertising were limited to established drugs then this problem could be lessened.

Many new drugs are significantly more expensive than their earlier counterparts while overall spending on pharmaceuticals has skyrocketed. The NIHCM Foundation indicated that the increase in spending on prescription drugs in the US in the year between 2000 and 2001 amounted to 17.1 percent and that this increase was made up of a rise in number of prescriptions (6.7 percent), price increases (6.3 percent), and switches to higher priced drugs (4.1 percent) (GAO 2002, 5).

These expenditures drain health care resources. The conclusion of many is that we cannot afford it. The NIHCM Foundation (2000, 15) concluded cautiously that:

> Our results do not address the affect [sic] of DTC ads on the public's health. But they add to the growing circumstantial evidence that such ads are one element – and perhaps an increasingly important one – in the recent trend to the expanded use of newer prescription drugs and the resultant increased overall spending on pharmaceuticals.

The New Zealand family physician authors conclude that the increase in prescription drug costs resulting from DTCA will be unsustainable and will create distortions in drug spending that will create an intolerable burden on the remaining health system (Toop et al. 2003, 19). They pointed out that the commentators in the European debate "predicted an unsustainable spiral of health care spending." The *Wall Street Journal* reported in a series of articles in March 2002 on the difficulties health insurers are experiencing with rising drug expenditures, which have risen more than 16 percent annually for large employers since 1997, and the resulting threat to the health care system. For example, General Motors spent US$55 million in 2001 on one drug, Prilosec, a heartburn drug, for GM employees – thirteen times more than it would have spent on a leading generic (Burton 2002).

The New Zealand report reaches ten conclusions (Toop et al. 2003, vi-viii):

1. DTCA is a very effective marketing strategy and is growing exponentially;
2. Consumer Education: DTCA does not provide objective information on risks, benefits and options to assist patients to participate in healthcare decisions;
3. Consumer information: Consumers need greater access to reliable independent information on prescription medicines;
4. DTCA has a negative effect on health funding which may create inequity in resource allocation;
5. DTCA has a negative effect upon the patient-clinician relationship.
6. DTCA compromises patient safety;
7. DTCA promotes the medicalisation of normal health and ageing processes;
8. There is increasing opposition to DTCA internationally amongst consumer and professional groups;

9.   There is increasing opposition to DTCA in New Zealand amongst consumer
      and professional groups;
10. DTCA cannot be controlled by either central or self regulation.

Their final conclusion is that DTCA cannot be controlled either by the state or through industry self-regulation. Although the United States has a good regulatory system, it still can't control the industry: "Even though they have much stricter regulations, there is still significant political, professional and consumer concern over the failure of their regulatory framework to prevent misleading advertising and the negative financial and health effects of DTCA" (id., viii). Later in the report they comment that, "it is clear that even a well-funded formal regulatory system such as the FDA in the United States cannot effectively control DTCA" (id., 45). The New Zealand Report concludes that the New Zealand government should ban all DTCA.

CONCLUSION

Advertising is directed to the minds of its targets – their attitudes, fears, hopes and beliefs. Motivators are placed in the ads to stimulate prescribing on the part of physicians and desire on the part of patient/consumers. The myths that make up drug ads sell lifestyles, legitimacy of the product, health through drugs. Direct-to-consumer advertising gives the pharmaceutical industry access to the minds of patients, unmediated by physicians. Physicians too are subject to the myths in drug advertising. In the name of education the industry is targeting patients, to direct their consumption. Ads are designed to direct people to consume – to have their desires satisfied through drugs.

Public health is a collective good. This premise of public health as a collective good carries with it an entitlement to action by the state. The Beijing Platform for Action, coming out of the United Nations Fourth World Conference on Women, called on national governments to "Take all appropriate measures to eliminate harmful, medically unnecessary or coercive medical interventions, as well as inappropriate medication and over-medication of women, and ensure that all women are fully informed of their options, including likely benefits and potential side-effects, by properly trained personnel" (United Nations 1995, para 106h). The Working Group on Women and Health Protection in Canada cited this part of the Platform for Action in calling for a ban on direct-to-consumer advertising (1999). The public is entitled to health information that meets the highest standards, to a health care system that is subjected to public health standards, and to free expression that is held to a standard of human dignity and equality in a free and democratic society. Without action by the state to control direct-to-consumer advertising significant adverse effects will be experienced by the public health system.

NOTES

I am grateful to Sarah Viau for her research assistance on this paper. Earlier versions of this chapter were presented at the 28th International Congress on Law and Mental Health in

Sydney, Australia (2003) and at the Health Law Teachers Conference of the American Society of Law, Medicine and Ethics in Wilmington, Delaware (2003) and I would like to thank the organizers and participants at those conferences.

1. 21 CFR. 202.
2. *Food and Drug Regulations*, C.01.044.
3. See: PAAB (Pharmaceutical Advertising Advisory Board), and Therapeutic Products Directorate (formerly Drugs Directorate). 1996. Roles and consultation related to advertising review, January 11, and document cited in footnote 15 of Part I.
4. *Federal Food, Drug and Cosmetic Act*, 502(n).
5. *Hollis v. Dow Corning Corp.*, [1995] 4 SCR 634.
6. *Buchan v. Ortho Pharmaceuticals (Canada) Ltd* (1986), 25 DLR (4th) 658 (Ont. CA).
7. *Illinois ex rel. Madigan v. Telemarketing Associates, Inc.*, 538 U.S. 600, 123 S. Ct. 1829 (5 May 2003) per Justice Ruth Bader Ginsburg for the unanimous court.

REFERENCES

Avorn, J., M. Chen, and R. Hartley. 1982. Scientific versus commercial sources of influence of the prescribing behavior of physicians. *American Journal of Medicine* 93: 4-8.

Barthes, R. 1972. *Mythologies*. New York: Hill and Wang.

Bell, R.A., R.L. Kravitz, and M.S. Wilkes. 1999. Direct-to-consumer prescription drug advertising and the public. *Journal of General Internal Medicine* 14(11): 651-7.

Bell, R.A., M.S. Wilkes, and R.L. Kravitz. 1999. Advertising-induced prescription drug requests: Patients' anticipated reactions to a physician who refuses. *Journal of Family Practice* 48(6): 446-52.

———. 2000. The educational value of consumer-targeted prescription drug advertising. *Journal of Family Practice* 49(12): 1092-8.

Burton, B. 2004. Drug industry to fight New Zealand's move to ban direct to consumer advertising. *British Medical Journal* 328: 1036.

Burton, T.M. 2002. Backlash rises against flashy ads for prescription pharmaceuticals. *Wall Street Journal*, 13 March.

Cockburn, J., and S. Pit. 1997. Prescribing behaviour in clinical practice: Patients' expectations and doctors' perceptions of patients' expectations – A questionnaire study. *British Medical Journal* 315(707): 520-3.

FDA (Food and Drug Administration, United States). 1997. *Draft guidance for industry: Consumer-directed broadcast advertisements*. Washington.

———. 1999. *Guidance for industry: Consumer-directed broadcast advertisements*. Washington. The updated version of the guidance is available at: <http://www.fda.gov.cder/guidance.1804fnl.hml> (Last accessed: 4 July 2003).

Fidler, D.P. 2002. A globalized theory of public health law. *Journal of Law, Medicine & Ethics* 30: 150-61.

GAO (General Accounting Office, United States). 2002. *Prescription drugs: FDA oversight of direct-to-consumer advertising has limitations* (October 2002) GAO-03-177.

Gilhooley, M. 2003. Drug regulation and the Constitution after *Western States*. *University of Richmond Law Review* 37: 901-33.

Goldman, R. 1992. *Reading ads socially*. London: Routledge.

Goldman, R., and S. Papson. 1996. *Sign wars*. New York: Guilford Press.

Goldman, R., and M. Montagne. 1986. Marketing mind mechanics: Decoding antidepressant drug advertisements. *Social Science & Medicine* 22: 1047-58.

HAI (Health Action International). 2002. *European Parliament soundly rejects move towards direct-to-consumer advertising: HAI Europe applauds Parliament's decision to protect public health, not industry's interests* (press release) 24 October. Available at: <http://www.haiweb.org/campaign/DTCA/pressreleaseEUrejectsDTCA.html> (Last accessed: 15 May 2003).

Healy, D. 1997. *The antidepressant era*. Cambridge: Harvard University Press.

———. 2002. *The creation of psychopharmacology*. Cambridge: Harvard University Press.

IMS Health Integrated Promotional Services. 2002. *Total U.S. promotional spending by type, 2001.* Fairfield, IMS Health. Available at: <http://www.imshealth.com/public/structu> (Last accessed: 17 July 2002).

Lock, M. 1993. *Encounters with aging: Mythologies of menopause in Japan and North America.* Berkeley: University of California Press.

Maguire, P. 1999. How direct-to-consumer advertising is putting the squeeze on physicians. *American College of Physicians – American Society of Internal Medicine Observer.* (March). Available at: <http://www.acponline.org/journals/news/mar99/squeeze.htm> (Last accessed: 23 February 2005).

Medawar, C. 1996 and ongoing. *The antidepressant web.* Available at: <http://www.socialaudit.org.uk/1.4.html> (Last accessed: 18 April 2005).

———. 2001. Health, pharma and the EU. A briefing for members of the European Parliament on direct-to-consumer drug promotion. Available at: <http://www.socialaudit.org.uk/5111-005.htm> (Last accessed: 13 January 2004).

Mintzes, B. 2001. *An assessment of the health system impacts of direct-to-consumer advertising of prescription medicines (DTCA). Volume II: Literature review.* University of British Columbia, Centre for Health Services and Policy Research, Research Report. Vancouver: Centre for Health Services and Policy Research. Available at: <http://www.chspr.ubc.ca/hpru/pdf/dtca-v2-litreview.pdf> (Last accessed: 10 May 2005).

Mintzes, B., M.L. Barer, R.L. Kravitz, L. Bassett, J. Lexchin, A. Kazanjian, R.G. Evans, R. Pan, and S.A. Marion. 2003. How does direct-to-consumer advertising (DTCA) affect prescribing? A survey in primary care environments with and without legal DTCA. *Canadian Medical Association Journal* 169: 405-12.

Mitka, M. 2003. Survey suggesting that prescription drug ads help public is met with skepticism. *Journal of the American Medical Association* 289(7): 827-8.

Moynihan, R., I. Heath, and D. Henry. 2002. Selling sickness: The pharmaceutical industry and disease mongering. *British Medical Journal* 324: 886-91.

Moynihan, R., and R. Smith. 2002. Editorial: Too much medicine? *British Medical Journal* 324: 859-60.

NIHCM (National Institute for Health Care Management) Foundation. 2000. *Prescription drugs and mass media advertising.* Washington.

———. 2002. *Changing patterns of pharmaceutical innovation.* Washington, DC: NIHCM.

Palumbo, F.B. and C.D. Mullins. 2002. The development of direct-to-consumer prescription drug advertising regulation. *Food and Drug Law Journal* 57: 423-43.

Pear, R. 1999. Drug concerns get FDA reprimands over advertising. *New York Times,* 28 March.

Peppin, P., and E. Carty. 2001a. Innovation, myths and equality: Constructing drug knowledge in research and advertising. *Sydney Law Review* 23: 543-76.

———. 2001b. Semiotics, stereotypes, and women's health: Signifying inequality in drug advertising. *Canadian Journal of Women and the Law.* 13: 326-60.

———. 2003. Signs of inequality: Constructing disability in antidepressant drug advertising. *Health Law Journal* 161-94.

Pinto, M.B., J.K. Pinto, and J.C. Barer. 1998. The impact of pharmaceutical direct advertising: Opportunities and obstructions. *Health Marketing Quarterly* 15: 89-101.

Reinhardt, U. 2001. Perspectives on the pharmaceutical industry. *Health Affairs* 20: 136.

Relman, A.S., and M. Angell. 2002. How the drug industry distorts medicine and politics: America's other drug problem. *The New Republic,* 16 December, 27-41.

Rosenthal, M.B., E.R. Berndt, J.M. Donohue, R.G. Frank, and A.M. Epstein. 2002. Promotion of prescription drugs to consumers. *New England Journal of Medicine.* 346: 498-505.

Slaughter, E. 2001-02. Fifth annual survey, consumer reaction to DTC advertising of prescription medicines: DTC advertising and self-care – Honoring requests for prescription medicines. *Prevention Magazine Report* 18.

TPP (Therapeutic Products Programme, Canada). 1999. Direct-to-consumer advertising of prescription drugs. Discussion document. Ottawa: Health Canada (April 6).

———. 1996, updated 3 November, 2000. The distinction between advertising and other activities, 12 January. Available at: <http://www.hc-sc.gc.ca/hpfb-dgpsa/tpd-dpt/advsactv_e.html> (Last accessed: 20 December 2004).

Toogoolawa Consulting, 2002. *Report of a review of advertising therapeutic products in Australia and New Zealand.* Commonwealth of Australia.

Toop, L., D. Richards, T. Dowell, M. Tilyard, T. Fraser, and B. Arroll, 2003. *Report to the Minister of Health supporting the case for a ban on DTCA, direct to consumer advertising of prescription drugs in New Zealand: For health or for profit?* New Zealand Department of Family Practice. Available at: <http://www.haiweb.org/campaign/DTCA/DTCAinNZcaseforaban2003.pdf> (Last accessed: 3 March 2003).

Williamson, J. 1978. *Decoding advertisements.* London: Marion Boyars.

Wolfe, S. 2002. Direct-to-consumer advertising – Education or emotion promotion? *New England Journal of Medicine* 346(7): 524-6.

Working Group on Women and Health Protection. 1999. *To do no harm: Direct-to-consumer advertising of prescription drugs.* Available at: <http://www.hc-sc.gc.ca/hpb-dgps/> (Last accessed: 28 January 2002).

CHAPTER NINE

TIMOTHY CAULFIELD & BARBARA VON TIGERSTROM

GLOBALIZATION AND BIOTECHNOLOGY POLICY

The challenges created by gene patents and cloning technologies

In the contemporary era, there are few areas of health policy that are untouched by the phenomena collectively referred to as globalization. More than ever before, national and supranational levels of regulation are intertwined, international forces affect domestic policy choices, and issues transcend boundaries, requiring collective responses. In the context of biotechnology, globalization adds further complexity to policy-making in an area that is already clouded by moral ambiguity, scientific uncertainty and rapid technological advances. This paper examines some of the implications of globalization for the development of domestic and international biotechnology policy, focusing on two areas: human cloning laws and genetic patent policy. It is meant to give the reader a flavor of the issues created by globalization pressures rather than to provide a comprehensive analysis of this complex area. Indeed, as will be clear, the complexity of these issues is such that it is difficult to make broad generalizations about the impact of globalization. Different aspects of globalization have implications for different areas of policy. The two examples discussed here are intended to illustrate the range of policy challenges that can arise.

In attempting an analysis like this, one is immediately confronted with the challenge of coming to grips with what globalization is, particularly given that this multifaceted phenomenon is understood in quite different ways by different people. This is not the place to attempt a definitive or comprehensive definition of globalization. Nevertheless, it may be useful to start by identifying some of the key aspects of globalization that are most relevant here. Globalization has been defined as "the expanding scale, growing magnitude, speeding up and deepening impact of interregional flows and patterns of social interaction" (Held and McGrew 2003, 4). It has spatial, temporal and cognitive dimensions, and all of these affect health and health policy, directly or indirectly (Lee, Fustukian and Buse 2003, 6-10; Held and McGrew 2003, 3-4; Lee 2003, ch 3-5). Developments in communications technology and the availability of faster, more accessible transport have blurred spatial boundaries (the spatial dimension), accelerated interactions (the temporal

B. Bennett & G.F. Tomossy (eds.), Globalization and Health: Challenges for Health Law and Bioethics,
129-149.
© 2006 *Springer. Printed in the Netherlands.*

dimension), and increased our awareness of events in distant parts of the globe (the cognitive dimension). National governments are increasingly faced with social problems and issues that transcend national and even regional boundaries, and demand common policy solutions.

A central aspect of globalization is the increasing degree of economic integration, the "global marketplace" characterized by large volumes of international trade, increasing mobility of labor and capital, and the growing economic power and global networks of transnational enterprises. The particular model of economic globalization that currently predominates – focused on liberalizing international trade and creating conditions for the global operation of market forces with a minimum of state interference – has important implications for the role of national governments and their capacity to promote other social objectives. At the same time, economic globalization creates incentives to develop uniform policies on issues affecting economic activity.

In many areas, the ideal would be for the international community to coordinate regulatory policy as much as possible. Without such coordination, there is likely to be a degree of corporate forum shopping and we will be unable to respond rapidly to emerging ethical, social and legal concerns. However, for many of the emerging policy issues, international cooperation, though a laudable goal, seems largely unattainable, at least in the short term. There remains a high degree of uncertainty in the international community about even basic issues, such as the application of the notion of human dignity in the context of reproductive cloning, the moral and legal status of the embryo, and the value, role and impact of gene patents. Of course, differences of opinion within countries may make it difficult to develop domestic laws. At the international level, however, the differing cultural and socio-political positions magnify the policy-making challenge. The cognitive dimension of globalization has made us more aware of issues and views in other parts of the world, but it has not necessarily led to cultural or normative convergence. In general, this is not necessarily a bad thing: most people would consider it a blessing that we can maintain a degree of cultural diversity in the face of the harmonizing forces of economic globalization. However, globalization has at the same time increased the desirability of harmonization in some areas of policy, because a common policy is the only way to prevent harmful practices that can more easily move across national boundaries.

This dilemma is illustrated by the example of cloning policy, which is discussed in the second part of this article. The first part discusses patent policy, particularly as it relates to gene patents, which presents somewhat different challenges. It illustrates the constraints on domestic policies that can result from aspects of economic globalization. Economic integration and the increasingly powerful international trade regime create pressures on governments to adopt harmonized policies. However, the common policy position that is emerging does not suit everyone's interests. Strong intellectual property protection is relied on by the biotechnology industry, but may interfere with public policy goals like the provision of affordable genetic testing.

There are, of course, many other biotechnology innovations that could be discussed, including genetically modified foods, xenotransplantation, and germ line therapy. This paper focuses on the two examples of human gene patents and human cloning to highlight different kinds of policy challenges that arise in the context of globalization.

PATENT POLICY

The economic dimension of globalization is particularly important in the area of patent policy. One truism of globalization is that countries are increasingly viewed as one player in a large, interconnected, international economy (UNEP 2002, 24). As such, policy makers motivated by economic development must now ask "how will this policy decision be perceived by international investors and markets?" In the fragile, knowledge-based sector of biotechnology, considerations of globalization seem particularly important. As suggested by Lester Thurow (2000, 21):

> The knowledge-based economy is fundamentally transforming the role of the nation-state. Instead of being a controller of economic events with in its borders, the nation-state is increasingly having to become a platform builder to attract global economic activity to locate within its borders.

Nowhere is the interplay between the forces of globalization and national policy more apparent than in the context of human gene patents. The economic realities of globalization add a layer of complexity that has the potential to frustrate the most worthy suggestions for patent reform. Moreover, they will intensify the already apparent conflict between the national health care and innovation agendas (see also: Caulfield 2003b; Gold 2002). At issue here are increasing economic integration and the harmonizing influence of developments in international economic law, in particular the World Trade Organization (WTO) agreements.

Though human gene patents have long been a source of social controversy (Andrews 2002; Caulfield, Gold and Cho 2000), there has never really been any doubt about the patentability of human genes – at least from the perspective of patent offices throughout the world – provided that the basic requirements of patentability are met.[1] Following the trend set by the influential 1980 US Supreme Court case of *Diamond v. Chakrabarty*,[2] if an inventor can isolate, purify and make useful a naturally occurring substance, even human DNA, it has the potential to be deemed patentable.[3]

The biotechnology industry is closely tied to patent policy. Few other industries rely on patent protection as much as this sector (Cook-Deegan and McCormack 2001; Straus 1998). It is not surprising, then, that a huge number of DNA patents have already been granted or applied for.[4] Because potential biotechnology products can take years to bring to market, investors believe they need strong Intellectual Property (IP) protection in order to provide a small sense of long-term security. Indeed, when a rough map of the human genome was completed in 2000, President Clinton and Prime Minister Blair made a joint announcement suggesting that the "gene map belongs to all" (Evenson 2000, A1). The statement was not meant to implicate patent policy. Nevertheless, the fear that the two leaders would seek to diminish intellectual property protection sent stocks in biotechnology companies tumbling (*Nature Biotechnology* 2000).

Though gene patents have long been perceived as a necessary component of the commercialization process, they have also been associated with many ethical, legal and social issues. The concerns are varied, including issues about commodification, human dignity and "biopiracy" (Caulfield, Gold and Cho 2000; Resnik 2001; Sarma 1999). However, the intensifying concerns about the impact of human gene patents on future health research and access to useful technologies seem to have been the impetus for the most recent policy recommendations calling for specific patent reform. In Canada, this patent dilemma was moved to the front of the policy making agenda by Myriad Genetics' decision to send "cease and desist" letters to provincial labs testing for the BRCA1/2 (breast cancer) mutations. Myriad Genetics, a US company, owns the patent on these genes and, therefore, has legal control over all testing. The Myriad test is quite expensive as compared to the testing process already being done in Canada, and the additional cost for public health authorities in Canada is perceived to be prohibitive.[5] In another recent example, the announcement that Genetic Technologies (GTG), an Australian company which owns broad patents on "non-coding" DNA, would be seeking millions of dollars in license fees from district health boards in New Zealand for their use of genetic testing raised serious concerns and has re-ignited debate about gene patents (Johnston 2003).[6]

Given the importance and profile of health care to most national and regional political agendas (Commission on the Future of Health Care in Canada 2002), it is hardly surprising that a perceived adverse impact on access, cost and sustainability would generate a desire to re-examine the role and value of patents. As noted by Lori B. Andrews (2002, 67), "Gene patents are undergoing increasing scrutiny because they are pivotal to the future of health care, and patent decisions around the globe have been most challenged and changed when health is at issue."

In the past few years many policy-making entities have provided suggestions on how to mediate the adverse effects of patents on the health care and research agendas. For example, in late 2002 the UK Nuffield Council on Bioethics recommended, *inter alia*, that patenting criteria (utility, novelty and non-obviousness) be more stringently applied, that patents over research tools be discouraged and that "compulsory licensing may be required to ensure reasonable licensing terms" (Nuffield Council on Bioethics 2002, ix).[7] In Canada, there are also examples of calls for patent reform. The Ontario government's Report to the Premiers (2002), which was adopted by all Canadian Premiers, recommends a clarification of patent criteria in relation to human genes, the exclusion of broad based genetic patents covering multiple uses, a clarification of the experimental and non-commercial exceptions and an expansion of the methods of medical treatment exclusion. More broadly, the Ontario Report (id., 15) suggests that:

> The goal of any patent reforms should be to uphold the beneficial aspects of patent law (e.g., encouraging research, invention and innovation) while ensuring a better balance between public and private interests with appropriate transparency and rigour.

The Australian Law Reform Commission (ALRC) recently completed a major inquiry on gene patenting to examine concerns about its impact on research and health care (ALRC 2004). It did not recommend any radical changes to the patent system, but suggested that the assessment of utility should be revisited, a new experimental use exception should be included in Australian patent legislation and the government should explore the invocation of Crown use provisions for the purposes of promoting health care and research.

Globalization, however, will make even moderate reforms challenging. First, existing international trade agreements, which both reflect and reinforce the phenomenon of economic globalization, have the potential to create legal barriers for those seeking to alter established patent rules. In particular, the *Agreement on Trade-Related Aspects of Intellectual Property Rights* (TRIPS) of the World Trade Organization, which came into effect in 1995 and currently involves 148 nations, was designed to harmonize and promote strong patent protection.[8] This agreement was built on the premise that strong intellectual property protection was needed to promote innovation and economic growth, and reduces the scope for unilateral patent reform. As such, it is unclear whether the recommendations made by entities like the Nuffield Council and the Ontario government would survive a challenge under TRIPS.[9] Australia's obligations under the TRIPS Agreement are repeatedly raised in connection with the options for reform discussed in the ALRC report (ALRC 2004).

It is reasonably clear that certain options that have been mooted would violate a state's obligations under the TRIPS Agreement: for example, to exclude a certain category of inventions – such as gene patents – would run afoul of Article 27(1), which requires that patents be available for inventions "in all fields of technology" (subject to the usual requirements for patentability). Shortening the term of patent protection would also be contrary to TRIPS obligations, since Article 33 prescribes a minimum period of protection. Other potential measures would appear to be permitted, but within certain limits. The provisions of TRIPS leave scope for exclusions from patentability for methods of treatment for humans or animals (Article 27(3)) and where preventing the commercial exploitation of an invention is necessary to protect *ordre public* or morality or to protect human life or health (Article 27(2)). It also allows states to provide limited exceptions to patent rights (Article 30). Granting compulsory licenses, a strategy that has been proposed to deal with gene patent issues, is permitted under Article 31, but only subject to certain conditions.[10] These provisions do allow some degree of flexibility, but their application and the exact scope of flexibility are uncertain.

The scope for member states to make policies to protect public health has been thoroughly canvassed in the context of pharmaceutical patents and the controversy over access to AIDS medication. Although the issues in that context are somewhat different from those relating to gene patents, there are some common threads, in particular concerns about the impact of patent protection on the costs of providing health care to national populations. In response to the acute difficulties of developing countries in obtaining access to AIDS drugs at an affordable cost, and their ensuing battles with the pharmaceutical industry and developed countries (in particular the US), a declaration on the subject was adopted at the Doha Ministerial Conference in 2001. The *Declaration on the TRIPS Agreement and Public Health* was the result of concerted lobbying by developing country parties (see Abbott 2002), and reaffirmed "the right of WTO Members to use, to the full, the provisions in the TRIPS Agreement, which provide flexibility" for the purpose of protecting public health.[11] In it, the Members "agree that the TRIPS Agreement does not and should not prevent Members from taking measures to protect public health" and "affirm that the Agreement can and should be interpreted in a manner supportive of WTO Members' right to protect public health" (paragraph 4). The Agreement is (in accordance with accepted principles) to be interpreted in light of its objectives and principles, which include the protection of public health (TRIPS Article 8(1)) and the "promotion of technological innovation...in a manner conducive to social and economic welfare" (TRIPS Article 7).

The Declaration should influence the interpretation of the TRIPS Agreement in any future dispute (Abbott 2002, 490ff), and stands as a clear statement of the determination of some governments not to allow trade obligations to unreasonably interfere with their capacity to protect the health of their people. To the extent that gene patenting raises concerns about accessibility and cost of treatments, it might be possible to invoke the Declaration in support of patent reforms. However, the Declaration does not create any new exceptions or reduce members' obligations in the area of patent protection, but rather merely reaffirms that some flexibility exists and can legitimately be used.[12] At best it suggests that a member state may be given some degree of benefit of the doubt if a dispute arises over a measure with an apparent public health purpose – and this in the context of efforts to address a large-scale public health emergency, so we cannot expect that calls for reform on gene patenting will be found any more compelling.

TRIPS obligations will therefore act as a constraint on policy in this area, given the strong pressures to comply. Signing on to the TRIPS Agreement is not optional for WTO members; if a state wants to be a member of the WTO it agrees to be bound by the WTO Multilateral Trade Agreements including the TRIPS Agreement. Opting out of the TRIPS Agreement is therefore not a real alternative for a national government that finds the Agreement's provisions too restrictive. Furthermore, the TRIPS Agreement is subject to the WTO dispute settlement mechanism which can result in the government having to change its policies or face WTO-authorized retaliatory measures for non-compliance. Arguably a state can decide not to comply – or to take the risk that its reforms will amount to non-compliance, where there is uncertainty – and face these penalties, but the economic and political cost of such a decision may be unacceptably high. At a minimum, existing international obligations add an element of risk and uncertainty to the policy-making process that may discourage attempts at reform.

As if this were not enough, states may also face pressure even to adopt more stringent patent protection than is required under TRIPS – a so-called "TRIPS-plus" approach – as was the case regarding pharmaceutical patents and the controversy over AIDS medication. So, for example, while compulsory licensing is permitted, at least under certain conditions under TRIPS, a TRIPS-plus approach might well be hostile to any form of compulsory licensing. TRIPS-plus policies may be adopted under political and economic pressure from trading partners. These pressures are intensified by economic globalization, including increasing integration and the influence of global economic powers. Powerful countries, such as the United States, may also require their trading partners to agree to TRIPS-plus protection in bilateral trade agreements. In addition, there may soon be the possibility of "non-violation" complaints regarding TRIPS in the WTO dispute settlement system, which would allow one WTO member to complain, in some circumstances, about the policies of another member even if those policies strictly complied with its TRIPS obligations.[13] This could potentially have the effect of enforcing TRIPS-plus protection despite the fact that it is not legally required. As a result, international legal obligations with respect to patent protection and the harmonizing effect of economic and political influence, can operate together to limit the policy options of national governments.

Globalization will also make reforms challenging in a second respect. Because intellectual property has become so closely tied to the broader political goal of economic development (Aoki 1998), any reform that has the potential to adversely impact the economic value of patents will meet a degree of resistance from both industry and those within government motivated by economic growth. Indeed, intellectual property protection is increasingly viewed as the economic foundation of the knowledge-based economy (Thurow 2000, 28). Juan Enriquez has gone so far as to suggest that a country or region's economic worth can be measured by the patents produced. "Patents are a good window (although not the only window) on who might triumph and who might lose over the course of the next two decades" (Enriquez 2001, 138). In biotechnology, patents are believed, rightly or not (Straus 1998; Buctuanon 2001),[14] to be an absolutely essential element of the innovation and commercialization process.[15] Janet Lambert, President of BIOTECanada, the national association of biotechnology researchers and practitioners, summarizes the industry position regarding biotechnology patents as follows:

> Intellectual property is often the most valued asset for biotechnology researchers and companies, particularly those who have yet to commercialize a product. The majority of Canadian biotechnology companies do not have revenues, are spending their capital on research, are small to medium sized businesses and are faced with increased international competition for funding and human resource skills. It is by ensuring Canada is internationally competitive in our protection of IP, that our voice as a society is heard in the vital social and ethical debates needed to define the boundaries of biotechnology.[16]   (*Canada NewsWire* 2002)

Globalization exacerbates this pressure to have "industry-friendly" patent policies, since firms and individuals are increasingly mobile, and may relocate to another jurisdiction if domestic policies are not favorable to their interests.

Finally, it is hard to overestimate the potential influence of US policy in this context. The harmonizing effect of globalization on patent policy does not reflect movement toward a global consensus, but rather a deeply contested position favored by economically and politically powerful actors. A large degree of the biotechnology industry is centered in the US, including most of the venture capital, much of the scientific activity and a large percentage of the consumer market.[17] This fact is reflected in the country's strong pro-patent approach to biotechnology. As noted by Elisa Buctuanon: "Being the world's major source of biotechnologies, it is but natural for the US system to secure the best protection of innovations generated" (Buctuanon 2001, 34). And, not surprisingly, most human gene patents have been filed in the US. Thomas, Hopkins and Brady (2002, 1186) found that of the human gene patents filed between 1996 and 1999 "62 percent were filed by organization in the United States, 20 percent in the European Union, and 10 percent in Japan." Likewise, a study by the OECD found that the US has generally, by far, the most biotechnology patents (OECD 2001). Of course, this reality has particular relevance to Canada, as any lessening of IP protection will stand in sharp contrast to the policies promoted by its largest trading partner.

More than anything, the forces of globalization intensify the policy tension between non-economic social goals, such as the desire to create a sustainable health care system, and the commercialization and innovation agenda. If, in fact, gene patents have an adverse impact on access to useful technologies and inappropriately drive up the cost of health care generally (Cho et al. 2003; Caulfield 2003b), governments must seek ways in which to strike an appropriate balance. Indeed, as intellectual property becomes an increasingly important part of the global economy, patents will concomitantly have a more significant impact on other social structures, such as health care systems and research institutions. Because social systems and needs can vary greatly between nations, a degree of flexibility seems appropriate.[18] Unfortunately, existing globalization forces, particularly those emanating from the US, allow for little policy experimentation and largely push in one direction – toward strong and uniform intellectual property laws. As asked by Lester Thurow (2000, 29), "A global system will have to allow for a diversity of economic positions and beliefs, but how is that system to come into existence?"

Given this state of affairs, countries with less economic influence – which includes Canada and, in reality, every country but the US – have a number of choices. First, they can move forward with patent reform and accept the probable, though perhaps short term, backlash from the global economy.[19] This may also include legal challenges under existing international agreements like TRIPS. The implementation of some form of compulsory licensing for genetic innovations, for example, may not be willingly accepted by the international community. Second, they can seek to make minor revisions that work within the existing patent rules and do little to alter the commercializability of useful technologies. The suggestion to apply existing patent criteria, particularly utility, more stringently would be an example of this approach.[20] Finally, Canada could work with the world community in the hope of developing an international patent policy that is more sensitive to the needs of, for instance, public health care goals. Unfortunately, as highlighted above, the existing globalization pressures makes this latter strategy the most challenging.

HUMAN CLONING POLICY

In the context of many biotechnologies, globalization also creates a clear need for a high degree of international cooperation. More than ever before, technologies and information can easily move between countries. The biotechnology industry is international in scope. And many emerging technologies implicate the entire world community. As such, if the regulatory goal is to contain or ensure the safe application of a given biotechnological innovation, some level of agreement between nations will be needed to control both the development and flow of the technologies. As noted by Allyn Taylor, the genetic revolution is "inherently international, necessitating collaborative multinational cooperation to promote global public health and to protect human rights while advancing scientific research and discovery" (Taylor 1999, 479).

Though it can be argued that compared to many emerging innovations, cloning is likely to have a relatively small impact on the human condition, it is a scientific development that has, rightly or not, generated a wave of national and international policymaking activity. As a result, the emerging attempts to regulate cloning and stem cell technologies stand as a good example of the challenges created by globalization in the context of morally contentious areas of research. If there is a desire to address the concerns related to the ethics of human reproductive cloning (see generally Daar 2003; Annas, Andrews and Isasi 2002; Caulfield 2003a; Pattinson 2002), international cooperation seems necessary. Without such cooperation, those who wish to proceed with the technology will merely need to locate in a jurisdiction without specific cloning laws.[21] This kind of forum shopping has the potential to undermine the regulatory goals of protecting human safety and dignity.[22]   In other words, the forces of globalization – the melding of economic forces, the sharing of information, etc. – arguably make international cooperation more necessary. As we shall see below, this understanding of the need for international cooperation was a primary motivator behind the United Nations' attempt to implement an international ban on reproductive cloning.

There are, however, numerous obstacles to making policies that have relevance to the entire world community. For example, the moral ambiguity that permeates the national discourse about the regulation of cloning technology is amplified on the international stage. Cultural, religious and political variations between nations make a consensus, at least on some topics, seem all but impossible.[23] There is, no doubt, an emerging agreement that reproductive cloning should not be allowed.[24] However, other technologies and procedures, such as "research cloning," are much more divisive.[25] And because reproductive and research cloning are often closely tied in the development of policy, consensus on how to move forward is difficult to achieve. Even among relatively culturally similar Western nations we see drastically different regulatory approaches emerging.

For example, within the European Union consensus has been difficult to obtain. Belgium and the UK allow the creation of embryos for research purposes. Finland, Greece, the Netherlands and Sweden, allow the procurement of stem cells from spare embryos. Austria, Denmark, Ireland and Spain all prohibit the creation of stem cell lines from embryos.[26] Sixteen European states are party to the *Convention on Human Rights and Biomedicine*, which prohibits the creation of embryos for research,[27] and thirteen are party to a Protocol to the Convention which prohibits cloning of human beings.[28] In the US, the states vary greatly in the approach to the regulation of research cloning. California and New Jersey have decided to explicitly allow cloning for research purposes while other states have already passed laws banning all forms of somatic cell nuclear transfer (Holden 2002; California Advisory Committee on Human Cloning 2002). And at the level of the US federal government, the President's Council on Bioethics' Report recently noted that a "moral consensus" did not exist on the topic of cloning and, as such, a moratorium, rather than a ban, was appropriate (The President's Council on Bioethics 2002). In Canada, there continues to be strong public support for the research and the government has taken what some consider a middle ground approach, allowing and funding embryonic stem cell research but prohibiting the creation of embryos for research purposes.[29]

This diversity of national and regional regulatory approaches makes the development of a single international position seem unlikely. Complicating things further, is the fact that few countries have adopted the same regulatory strategy (Pattinson and Caulfield 2004). In some countries, such as Canada and Australia, new legislation was developed that explicitly addresses that topic, employing a combination of legislative bans and regulation. In other countries, existing laws have been interpreted to apply, as is the case in Ireland. Also, many regulatory policies are in a state of flux. Indeed, many countries have laws which require mandatory legislative review (e.g. Canada's law must be reviewed in three years)[30] or are explicitly time limited (e.g. Israel's law is to apply for a period of five years).[31] More importantly, the policy justifications behind the laws often differ. As one of us has noted elsewhere: "Even where there is agreement as to the regulatory outcome, policy-makers should not confuse this with agreement on underlying ethical principles" (Pattinson and Caulfield 2004, 11). Of course, much of lack of consensus relates primarily to research cloning and research involving embryos as opposed to reproductive cloning. But the disagreements run deep. As noted by Monachello (2003, 18): "Although most nations agree and have passed laws banning cloning to produce children, a world wide consensus on the topic of cloning for biomedical research seems virtually impossible."

In addition, it should not be forgotten that though many of the Western nations with well developed research infrastructure have cloning and stem cell policies, most countries in the world have no relevant laws or policies (Bonnicksen 2002). For many countries, the creation of biotechnology policy is far from a national priority. As such, in a significant portion of the world, particularly in developing nations, there is little to stop those wishing to proceed with the development of controversial technologies. For an international approach to be effective, the domestic laws in all relevant jurisdictions will need to be developed to, at least, some minimum level.

The difficulty in developing international policy in this area is well illustrated by the frustrated attempts of the United Nations to produce a treaty on reproductive cloning.[32] In December 2001, the General Assembly established an Ad Hoc Committee to consider "the elaboration of an international convention against the reproductive cloning of human beings."[33] The Ad Hoc Committee met in February and March 2002 and work then continued within a working group of the General Assembly Sixth (Legal) Committee. Competing proposals were submitted, advocating different approaches. The first, exemplified by a proposal submitted by France and Germany, favored moving quickly to adopt a convention banning reproductive cloning, leaving the question of research or therapeutic cloning open for future discussion, and in the meantime, a matter of policy for individual states, which would be free to adopt more restrictive policies than required by the convention's ban on reproductive cloning.[34] The second, as seen in a proposal submitted by Spain, argued for a comprehensive ban on cloning, including for therapeutic purposes.[35] Despite attempts to propose compromise solutions,[36] some aspects of the disagreement seem intractable. Concerns about the clarity or practical effectiveness of the proposed ban, which were among those cited in favor of a comprehensive ban, might be possible to address.[37] However, the divergent approaches also reflect more fundamental differences about the nature and weight of ethical and moral concerns about the creation and use of human embryos for research or therapeutic purposes.

A draft convention was submitted by Costa Rica as a contribution to the UN negotiations.[38] It would require states to establish criminal offences relating to any type of human cloning, without restriction as to purpose.[39] The preamble to the draft convention stated that

the cloning of human beings, whether carried out on an experimental basis, in the context of fertility treatments or pre-implantation diagnosis, for tissue transplantation or for any other purpose whatsoever, is morally repugnant, unethical and contrary to respect for the person and constitutes a grave violation of fundamental human rights which cannot under any circumstances be justified or accepted.

The invocation of the universal language of human rights here masks the lack of consensus on the moral and legal basis for a comprehensive ban.

The Sixth Committee working group continued its work in the 58th session of the General Assembly in September and October 2003.[40] Participants expressed concern that limited progress had been made after two years of discussions and stressed the need for consensus, but there continued to be a divergence of views among participating delegations.[41] After a close vote (80-79) in November 2003, the Legal Committee recommended a two-year deferral on a General Assembly decision. This compromise was put forward and supported by most of the members of the Organization of the Islamic Conference (Ashwanden 2004). However, in January 2004, the General Assembly, which was under significant pressure from those countries wanting a more comprehensive ban, overturned the Legal Committee's recommendation and established a one year delay on the debate. Then, in October 2004, the General Assembly re-opened the debate, again with no apparent

compromise from either camp.[42] The debate ended, on 19 November, with the
General Assembly avoiding a vote on a comprehensive ban by agreeing to meet
in February 2005, to negotiate a non-binding statement that could guide the
development of national laws (Associated Press 2004; Reuters 2004). This was
realized in early March 2005, when the General Assembly adopted, by resolution,
the *United Nations Declaration on Human Cloning*.[43] This non-binding Declaration
calls upon UN Member States to "prohibit all forms of human cloning inasmuch
as they are incompatible with human dignity and the protection of human life," and
to adopt national legislation giving effect to this prohibition. The voting record
indicates that states continue to be deeply divided, however, and it is widely
recognized that adoption of the Declaration does not indicate that any meaningful
consensus has been reached.[44]

     This regulatory gridlock creates an interesting policy dilemma. Globalization
increases the need for an international approach to policy making in this context.
But, at the same time, an international approach to the regulation of all cloning
technology requires us to subvert the diversity of cultural positions on central moral
issues, most notably the moral and legal status of the embryo. All can agree that
reproductive cloning is currently an unsafe technology and, as such, policy makers
can safely justify a ban on that concern alone – even if other justifications, like the
impact of reproductive cloning on human dignity, remain hotly contested (Pattinson
2002). However, the concerns associated with other uses of cloning technology,
such as for research purposes, are based almost entirely on ethical principles which
may be closely tied to a given region's specific history or culture.[45] We could hope
for a degree of compromise and the finding of a middle ground. International
agreements are, after all, often built around compromise. But in the areas of stem
cell research and human cloning the values at play are either not amenable to
compromise (as with some views on the moral status of the fetus) or they remain
disputed or underdeveloped (as with the application of the notion of human dignity).
It is significant that some states reject even an approach which would explicitly
provide only for a "lowest common denominator" prohibition and leave the adoption
of more restrictive policies up to each state, as seen in the United Nations negotiations.
This is hardly the atmosphere to build thoughtful and lasting international
agreement.

     As such, unless we are willing to accept that the process of developing an
international agreement may, itself, become a globalization force that homogenizes
and simplifies complex and hotly debated ethical positions, we must allow room for
national variation and continued public debate. Admittedly, this will mean that in
jurisdictions that wish to immediately ban all forms of cloning, policy makers must
accept the reality that a degree of forum shopping may, inevitably, soften the
effectiveness of regional cloning regulations.[46]

     Finally, let us briefly consider the influence of global economics in this context.
Economic considerations, while not as dominant as in the area of patent policy, also
play a role in the policies of numerous countries. In fact, a number of countries have
developed policy based, in part, on the belief that the regulatory environment that

surrounds stem cell research will have a significant impact on the development of the biotechnology sector. For example, it is suggested that Singapore is:

> poised to become a beehive for biomedical research. This reflects a deliberate economic policy to diversify the electronics dominated manufacturing sector. The Singapore government has earmarked three billion dollars (US 1.6 billion) to promote research and development in the life sciences.   (Oriola 2002, 497)

To this end, the Singapore Bioethics Advisory Committee has recommended that a wide variety of embryonic stem cell research be permitted – including research cloning (Regnier and Knoppers 2002, *Stem Cell Week* 2002). At the same time, commentators have noted that the US's harsh stand on embryonic stem cell research and research cloning may hurt their research infrastructure.[47] Daar believes that a "byproduct of a [US] ban on scientific research would be a 'brain drain,' as talented and well-funded scientists leave the United States to establish research centers in nations with hospitable working environments" (Daar 2003, 570). In Canada, some have suggested that our more permissive regulatory environment may lead to a "brain gain" (Abraham 2002, A1).

CONCLUSION

In many areas of public policy, the various aspects of globalization add a layer of complexity to policy making, whether by narrowing the range of policy options, increasing the costs of certain policy decisions, or broadening the range of issues and implications that need to be examined before a decision can be made. As governments throughout the world struggle to develop policies to meet the unique social concerns created by biotechnology, appreciating the relevance of one of this era's defining phenomena, globalization, seems essential.

This article has outlined a few of the regulatory challenges created by globalization in the area of biotechnology. We see that globalization pressures can have the paradoxical effects of creating a need for international cooperation and making that cooperation difficult to obtain. Given the momentum of the international biotechnology industry, globalization will undoubtedly make it more difficult to drastically alter existing patent policies, even in the face of mounting evidence that such change may be required. Likewise, globalization forces will make it more difficult to regulate the development and use of controversial innovations, such as human cloning technology. But because of diverse moral and socio-political positions, building the international consensus necessary to effectively regulate the area seems all but impossible.

The two examples discussed in this paper demonstrate the complexity of coming to terms with the impact of globalization on biotechnology policy. Globalization is not a single unified phenomenon but rather a complex set of interconnected developments which do not necessarily have the same effects in every context. We have seen that there are some common themes that emerge: for example, in both gene patenting and cloning policy, the fact that individuals and firms are increasingly mobile creates pressures for policy harmonization. However, in the first case of patent policy, other aspects of globalization – in particular the impact of the WTO regime – are also pushing governments toward adopting a uniform position, while many of them feel that the common policy position is not appropriate or desirable. In the second case of cloning policy, a uniform policy at least on some points *is* desirable, and globalization makes it more desirable, but has not made it more achievable. One of the lessons that can be taken from this is that there can be no simple template for understanding and addressing the implications of globalization for biotechnology policy. This is an area that will continue to evolve and challenge policy makers in new and complex ways in the years to come.

NOTES

We would like to thank Cathy Jo Kickman, Lori Sheremeta, Nina Hawkins, Katherine Ellena and Alison Heath for valuable assistance and Genome Prairie, the Stem Cell Network and the Alberta Heritage Foundation for Medical Research for their continued support. This chapter was presented at the 28th International Congress on Law and Mental Health in Sydney, Australia (2003) and builds on the following article: Timothy Caulfield, "Gene Patents, Human Clones and Biotechnology Policy: A Comment on the Challenges Created by Globalization" (2003) 41 Alberta Law Review 713.

1. The essential requirements are utility, novelty and non-obviousness. If these requirements are not met, the patent may be refused or revoked. For example, in 2004, the European Patent Office decided to revoke two of Myriad Genetics' breast cancer gene patents after they were successfully opposed on this basis (European Patent Office 2004; Ghosh 2004).
2. 447 US 303 (1980). See generally Andrews and Nelkins (2001).
3. See Walsh (2002, 159) where it is noted: "There has been an ever increasing expansion of intellectual property protections, especially since the Second World War, and, more recently, rapid change in IPRs [intellectual property rights] in relations to biotechnology and what the public expects." Even products of nature are patentable "even though the novelty of human ingenuity involved might simply be purification or isolation from a plant or animal raw material" (id., 160).
4. As of the year 2000, over 25,000 DNA-based patents have been issued in the US alone (Cook-Deegan and McCormack 2001, 217). See also Thomas, Hopkins and Brady (2002, 1185) who suggests that "between 1996 and 1999, 6786 patents were filed" in the US, Europe and Japan.
5. Myriad charges more than CND $3800 per test (Benzie 2001). See also Gold, Caulfield and Ray (2002) and Hahn (2003) for a discussion of a similar controversy surrounding the patenting and commercialization of the Canavan gene.
6. The GTG non-coding patents are said to be so "wide-ranging" that they could be seen to apply to "virtually any genetic testing" (Johnston 2003). GTG initially sought a NZ$10 million fee for signing and waiving past infringements, and $2 million per year for a national licence that would cover all human diagnostic testing at public institutions. After protests by the Auckland District Health Board (ADHB), this amount was reduced. However, the ADHB has recently initiated legal action against GTG challenging its claims (*Scoop* 2004).
7. See also the recommendation of the UK Royal Society: "We recommend that governments further facilitate compulsory licensing and application of competition law in situations where single or multiple patents do, on balance, unreasonably affect use and development of inventions" (2003, para 3.19).
8. TRIPS was, to some degree, a direct result of the US trade policy and the US government "taking a very hard line against global piracy and pressuring foreign governments with weak patent laws to strengthen them" (Lehman 2002, 2).
9. See also Andrews (2002, 76), who notes that: "Article 8 of TRIPS allows governments to take public health concerns into consideration with their national intellectual property laws." However, given the

aggressive position of US on intellectual property issues, such an approach could face a challenge. See also Buctuanon (2001, 34): "the US influence over IPR [intellectual property rights] extends to the Trade Related Aspects of Intellectual Property (TRIPS) issues in the WTO because of the enormous stake of American firms on biotechnology products sold in the world market."

10. For example, there is a requirement of prior negotiation with the patent-holder (which can be waived in the case of a "national emergency or other circumstances of extreme urgency"), payment of adequate remuneration, and limits on scope and duration. In Caulfield, Gold and Ray (2002) it is argued that it is theoretically possible to implement a mandatory licensing scheme under TRIPS.

11. Declaration on the TRIPS Agreement and Public Health, para. 4. Examples of the "flexibilities" are listed in paragraph 5 and include the right to grant compulsory licences, to determine the grounds on which they are granted and to determine what constitutes a national emergency or circumstances of extreme urgency for the purposes of the compulsory licence provision.

12. It has since been supplemented by a decision of the WTO General Council which does provide a limited, temporary waiver of TRIPS obligations to allow for export production of pharmaceuticals under compulsory licence in cases of extreme urgency: Decision on implementation of paragraph 6 of the *Doha Declaration on the TRIPS Agreement and Public Health*, WT/L/540 (30 August 2003). This would not be helpful in the context of gene patenting policy and reinforces the point that the Doha Declaration does not, on its own, diminish legal obligations under TRIPS.

13. The WTO dispute settlement system is unusual in that it permits proceedings to be taken where another member's polices or practices have "nullified or impaired" a trade benefit, even if they do not violate the member's obligations under a WTO agreement. This can happen where a benefit reasonably expected by a trading partner has been removed or reduced. To date there has been a moratorium on these non-violation complaints regarding the TRIPS Agreement, but the moratorium is due to expire in July 2005 and the United States is strongly opposing its renewal (TRIPS Council 2004).

14. It is important to note that though this is the perception in industry, there remains little evidence to support the claim. As noted by Gold et al. (2002, 327): "Despite the assumption within intellectual property systems that they are necessary to encourage research and development, there is only a modest body of empirical evidence to support this in the biotechnology industry." See also Scherer (2002, 1363): "Offering patent patents rights cannot motivate what has already been done, it can only confer windfalls for past investments made without the clear expectation that patent rights could be secured."

15. See generally Straus (1998) and Buctuanon (2001, 34), where Buctuanon reports on a survey of biotechnology companies. "[T]he Intellectual Property Rights (IPR) regime emerged as the fourth crucial factor identified by the companies. This is because the risks and costs inherent in the development of biotechnology products are so high that the ability of a country to secure property interests (products, processes and know-how) becomes an incentive that will encourage firms to invest more in R&D."

16. We offer this quote merely to highlight the perception of the biotechnology industry. The reaction of the biotechnology community to the Harvard Mouse decision stands as a good example of how industry and policy makers may respond to patent reform. In December of 2002, the Supreme Court of Canada held that higher life forms are not patentable under Canadian patent law (*Harvard College v. Canada (Commissioner of Patents)*, [2002] 4 SCR 45). Even though patents on higher life forms arguably play a very small role in the broader biotech industry, the decision elicited a strong response. See Gervais 2003: "While the technical impact of the decision will likely be minor, the negative effect on future biotechnology could be huge." See also *Canada NewsWire* 2002, where Lambert suggests: "Today's decision destroys our Canadian infrastructure of knowledge and innovation, creates an even greater brain drain, and we will lose our place at the world table in influencing how and where society accepts this technology." See also Abraham (2002).

17. As noted by Buctuanon (2001, 26): "it appears that although the flow of biotechnology across national boarders has grown, increasingly in recent years, there is a tendency for this to agglomerate in developed countries, particularly the US, where the socio-economic and politico-institutional environment facilitates their development and commercial exploitation." Later, the author notes that

"A close examination on the direction of firms in the forming strategic R&D partnerships reveals that they are headed toward the US for insourcing biotechnologies" (ibid. at 29).

18. Of course, existing intellectual property rights create far greater challenges for the developing world. Though a discussion of this issue is beyond the scope of this paper, it is worth noting that a growing "genomics divide" is emerging. "[H]ealth research is disproportionately directed to the developed world, leading to the so-called 10/90 gap – 10 percent of the world's population accounts for 90 percent of research expenditure dedicated to health" (Dowdeswell, Daar and Singer 2003, 2-3). See also Davidge and Saner (2003, 9): "Signing on to TRIPS and thereby adopting a Western-style intellectual property regime greatly increases the cost of new technologies. TRIPS consolidates the position of technological leaders but does little spread the wealth."

19. One can only guess what the economic fallout out of drastic patent reform would be. Many suggested that the Harvard mouse decision (*Canada Newswire* 2002) would have a dramatic impact. To our knowledge, this has yet to materialize. However, a reform that would have more obvious ramifications to the commercialization process, such as compulsory licencing, may elicit a more substantial response from industry. Nevertheless, policy makers may feel that alleviating the potential adverse effects of gene patents on the cost of the health care system would off set the economic loss created by patent reform and therefore, in the aggregate, patent reform is justified. Clearly, more empirical evidence is needed to inform this type of policy making.

20. See, for example, Nuffield Council on Bioethics 2002. See also, the USPTO 2001 utility guidelines (Federal Registry, Vol 66, No. 4, January 5, 2001) and, for a discussion of other possible patent reforms see Sheremeta and Gold 2003.

21. It has also been suggested that an international ban on human cloning could never be truly effective because it would merely drive the research underground and to jurisdictions with little oversight of human research. Indeed, some commentators believe that a ban will "only encourage development of irresponsible and ill conceived research agenda" (Daar 2003, 571). As such, Daar advocates the development of a regulatory scheme that allows the technology to be developed and applied in a responsible manner.

22. Concern for human safety is largely viewed as a logical justification for the regulation of human reproductive cloning. See, for example, National Bioethics Advisory Commission 1997; Giles and Knight 2003. At the current time, however, there remains a great deal of disagreement about the role of human dignity in this context. Nevertheless, the protection of human dignity is an often used justification for cloning laws. See generally Beyleveld and Brownsword 1998; Malby 2002; Wright 2000.

23. Of course, in the context of embryonic stem cell research and research cloning, a great deal of the variation in approaches can be traced to views on the moral and legal status of the embryo. For example, among the major religions we see vastly different positions (Evans 2002a; 2002b).

24. For example, a 2002 Gallup poll found that 90 percent of Americans disapprove of cloning that is designed specifically to result in the birth of a human being (Saad 2002).

25. "Research cloning," also referred to as non-reproductive human cloning and therapeutic cloning, involves the use of somatic cell nuclear transfer for the purpose of cloning an "embryo" for research purposes. It is hoped that this technique could be used to create cell lines for research purposes or, even, tissue for transplantation (Daar and Sheremeta 2002).

26. For a discussion and review of national and international cloning policies see Pattinson and Caulfield 2004. See also Meldolesi (2003, 588). To add more confusion to the European scene, Germany prohibits embryonic stem cell research but allows the importation of human embryonic stem cell lines. See also Wertz, Regnier and Knoppers 2003. The various approaches to the regulation of "research cloning" are discussed in Caulfield (2003c).

27. *Convention for the Protection of Human Rights and Dignity of the Human Being with regard to the Application of Biology and Medicine (Convention on Human Rights and Biomedicine)*, 4 April 1997, ETS No. 164, article 18(2). At the time of writing, thirty-one states have signed the treaty, but only sixteen have ratified.

28. *Additional Protocol to the Convention for the Protection of Human Rights and Dignity of the Human Being with regard to the Application of Biology and Medicine (Convention on Human Rights and Biomedicine)*, 12 January 1998, ETS No. 168. The Protocol prohibits, article 1(1), "[a]ny intervention

seeking to create a human being genetically identical to another human being, whether living or dead" but does not define "human being" thus leaving the exact scope of the prohibition up to individual state parties. Twenty-nine states have signed, and thirteen ratified.

29. See the *Act Respecting Assisted Human Reproduction and Related Research (Assisted Human Reproduction Act)*. This law, which was proclaimed on March 29, 2004, prohibits reproductive and research cloning Section 5 states: "No person shall knowingly (a) create a human clone by using any technique, or transplant a human clone into a human being or into any non-human life form or artificial device; (b) create an in vitro embryo for any purpose other than creating a human being or improving or providing instruction in assisted reproduction procedures; (c) for the purpose of creating a human being, create an embryo from a cell or part of a cell taken from an embryo or foetus or transplant an embryo so created into a human being." For data regarding the views of Canadians, see Pollara (2002, 12): "Awareness and recall of stem cell research hovers over 60 percent with the vast majority of Canadians being at least somewhat supportive of the research. The number of people adamantly opposed has dropped five points to 13 percent of the population. Further, a vast majority of Canadians believe it is very or somewhat acceptable for the Government of Canada to be involved in supporting this type of research."

30. See *Assisted Human Reproduction Act*, section 70(1): "The administration of this Act shall, within three years after the coming into force of section 21, be reviewed by any committee of the Senate, the House of Commons or both Houses of Parliament that may be designated or established for that purpose."

31. Law 575 – 1999, Prohibition of Genetic Intervention (Human Cloning and Genetic Manipulation of Reproductive Cells).

32. For further discussion, see *Nature* (2002, 1331) where it is noted that a UN ban on research cloning would "set a precedent for global restrictions on basic research."

33. UN General Assembly, Res. 56/93, 12 December 2001, at para 1.

34. UN General Assembly, *International Convention Against the Reproductive Cloning of Human Beings: Report of the Working Group*, UN Doc. A/C.6/57/L.4, 30 September 2002, Annex I, Parts 1 and 2.

35. Ibid., Annex I, Part 3.

36. See ibid., Annex I, Part 5; Annex II.

37. See ibid., Annex I, Part 3; Annex II.

38. *Draft international convention on the prohibition of all forms of human cloning*, Annex I to the Letter dated 2 April 2003 from the Permanent Representative of Costa Rica to the United Nations addressed to the Secretary-General, UN Doc. A/58/73, 17 April 2003.

39. Ibid., articles 2, 3.

40. *UN General Assembly, International Convention Against the Reproductive Cloning of Human Beings: Report of the Working Group*, UN Doc. A/C.6/58/L.9, 3 October 2003.

41. Ibid., Annex II, Informal summary of the general discussion in the Working Group, prepared by the Chairman.

42. Ad hoc Committee on an International Convention Against the Reproductive Cloning of Human Beings, Update 15 November 2004. Available at: <http://www.un.org/law/cloning> (Last accessed: 2 May 2005).

43. UN General Assembly, Res. 59/280, 8 March 2005, Annex.

44. UN, Press Release GA/10333, 8 March 2005. There were 84 votes in favour of the resolution, 34 votes against, and 37 abstentions. In statements after the vote, many states expressed regret at the failure to reach consensus and the lack of clarity in the Declaration.

45. For example, Ireland's restrictions on research involving human embryos is obviously informed by the country's strong Catholic tradition. Likewise, in Israel, the government's support of embryonic stem cell research and therapeutic cloning flows, in part, from the Jewish perspective on the status of the early fetus. See Gross and Ravitsky (2003, 250): "In contrast to the Catholic view, which treats the embryo as a person from the moment of conception, Jewish and Muslim traditions see an embryo that progressively acquires human status during embryonic development. According to the Orthodox Jewish view, genetic materials outside the uterus have no legal status because they are not considered part of a human being until implantation. Therefore, the status of the preimplantation embryo outside

the womb is comparable to that of gametes – namely, it should not be wasted but may be manipulated for therapeutic purposes." See also, Rabbi Dr. Yigal Shafran, who heads the Department of Jewish Medical Ethics in the Chief Rabbinate of Jerusalem, has suggested that "cloning is part of science's desire to benefit the human race" (Traubman 2002).

46. There are, of course, many reasons why a given jurisdiction may wish to pass laws that govern what happens within its borders, even though a national law may do little to stop the eventual proliferation of a controversial technology (Daar 2003). Nevertheless, the law would still stand as a symbolic statement of the formal position of the nation on the appropriateness of the technology.

47. In the US, federal funding will only be provided for stem cell research involving the limited number of stem cell lines that existed prior to the development of President Bush's policy. The federal government is also considering a ban on all forms of research involving human cloning, including "research cloning" (*Nature* 2002).

REFERENCES

Abbott, F.M. 2002. The Doha Declaration on the TRIPS Agreement and public health: Lighting a dark corner at the WTO. *Journal of International Economic Law* 5: 469-505.

Abraham, C. 2002. Bush's stem cell policy could mean brain gain for Canada. *Globe and Mail*, 11 August.

Andrews, L.B. 2002. The gene patent dilemma: Balancing commercial incentives with health needs. *Houston Journal of Health Law and Policy* 2: 65-106.

Andrews, L.B., and D. Nelkins. 2001. *Body bazaar: The market for human tissue in the biotechnology age*. New York: Crown Publishers.

Annas, G., L.B. Andrews, and R. Isasi. 2002. Protecting the endangered human: Toward an international treaty prohibiting cloning and inheritable alterations. *American Journal of Law and Medicine* 9: 151–78.

Aoki, K. 1998. Sovereignty and the globalization of intellectual property. *Indiana Journal of Global Legal Studies* 6: 11-58.

Aschwanden, C. 2004. UN to vote on cloning in one year, not two. *Bulletin of the World Health Organization* 82: 76.

Associated Press. 2004. UN gives up global cloning ban. 19 November.

ALRC (Australian Law Reform Commission). 2004. *Genes and ingenuity: Gene patenting and human health*. Sydney: ALRC.

Benzie, R. 2001. Ontario to defy US patents on cancer genes: Province will pay for $800 test, not $3,850 version by Myriad Genetic Laboratories Inc.: 'Share the benefits.' *National Post*, 20 September.

Beyleveld, D., and R. Brownsword. 1998. Human dignity, human rights, and human genetics. *Modern Law Review* 61: 661-80.

Bonnicksen, A. 2002. *Crafting a cloning policy: From Dolly to stem cells*. Washington DC: Georgetown University Press.

Buctuanon, E.M. 2001. Globalization of biotechnology. *New Genetics and Society* 20: 25-41.

California Advisory Committee on Human Cloning. 2002. *Report of the California Advisory Committee on Human Cloning*. Available at: <http://www.sfgate.com/chronicle/cloningreport> (Last accessed: 24 January 2005).

Canada NewsWire. 2002. BIOTECanada responds to Supreme Court decision on Harvard mouse case. 5 December. Available at: <http://www.newswire.ca/en/releases/archive/December2002/05/c0202.html> (Last accessed: 23 February 2005).

Caulfield, T. 2003a. Human cloning laws, human dignity and the poverty of the policy making dialogue. *BMC Medical Ethics* 4: 3-10.

———. 2003b. Sustainability and the balancing of the health care and innovation agenda: The commercialization of genetic research. *Saskatchewan Law Review* 66: 629-45.

———. 2003c. The regulation of embryonic stem cell research: A few observations on the international scene. *Health Law Journal*. 2003 Special Edition: 87-95.

Caulfield, T., B.M. Knoppers, R. Gold, L. Sheremeta, and P. Bridge. 2002. Genetic technologies, health care policy and the patent bargain. *Clinical Genetics* 63(1): 15-18.

Caulfield, T., R.,Gold and M. Cho. 2000. Patenting human genetic material: Refocusing the debate. *Nature Reviews Genetics* 1: 227-31.

Caulfield, T., R. Gold, and P. Ray. 2002. Gene patents and the standard of care. *Canadian Medical Association Journal* 167: 256-7.

Cho, M., S. Illangasekare, M. Weaver, D. Leonard, and J. Metz. 2003. Effects of patents and licences on the provision of clinical genetic testing services. *Journal of Molecular Diagnostics* 5: 3-8.

Commission on the Future of Health Care in Canada. 2002. *Building on values: The future of health care in Canada*. Ottawa: Government of Canada Publications.

Cook-Deegan, R.M., and S.J. McCormack. 2001. Patents, secrecy and DNA. *Science* 293: 217.

Daar, J. 2003. The prospects of human cloning. Improving nature or dooming the species. *Seton Hall Law Review* 33: 511-72.

Daar, A., and L. Sheremeta. 2002. The science of stem cells: Some implications for law and policy. *Health Law Review* 11: 5-9.

Davidge, A., and M. Saner. 2003. *Bridging troubled waters: Canada's role in connecting biotechnology to global human needs*. Ottawa: Institute on Governance.

Dowdeswell, E., A. Daar, and P. Singer. 2003. Bridging the genomics divide. *Global Goverance* 9: 1-6.

Enriquez, J. 2001. *As the future catches you*. New York: Crown Business.

European Patent Office. 2004. "Myriad/breast cancer" patent revoked after public hearing. Press release, 18 May.

Evans, J. 2002a. Religion and human cloning: An exploratory analysis of the first available opinion data. *Journal for the Scientific Study of Religion* 41: 747-58.

————. 2002b. Cloning Adam's rib: A primer on religious responses to cloning (A Report Prepared for the Pew Forum on Religion and Public Life, 2002). Available at: <http://www.pewforum.org/publications/reports/adamsrib.pdf> (Last accessed: 23 February 2005).

Evenson, B. 2000. Gene map belongs to all. *National Post*, 13 March.

Gervais, D. 2003. Eeek! The mouse confounds the house! *The Ottawa Citizen*, 14 December.

Ghosh, S. 2004. Myriad troubles facing gene patents. *Preclinica* 2(5): 300.

Giles, G., and J. Knight. 2003. Dolly's death leaves researchers woolly on clone ageing issue. *Nature* 421: 776.

Gold, E.R. 2002. Biotechnology patents: Strategies for meeting economic and ethical concerns. *Nature Genetics* 30: 359.

Gold, R., D. Castle, L. Cloutier, A. Daar, and P. Smith. 2002. Needed: Models of biotechnology intellectual property. *Trends in Biotechnology* 20(8): 327-9.

Gross, M., and V. Ravitsky. 2003. Israel: Bioethics in a Jewish-democratic state. *Cambridge Quarterly of Healthcare Ethics* 12: 247-55.

Hahn, L. 2003. Owning a piece of Jonathan. *Chicago*, May, 83.

Held, D., and A. McGrew. 2003. The great globalization debate. In: *The global transformations reader,* 2nd edition, eds. D. Held, and A. McGrew, 1-50. Cambridge: Polity Press.

Holden, C. 2002. California flashes a green light. *Science* 297: 2185.

Johnston, M. 2003. Aust firm slaps $10m bill on NZ's use of patented DNA test methods. *New Zealand Herald*, 12 August.

Lee, K. 2003. *Globalization and health: An introduction.* New York: Palgrave Macmillan.

Lee, K., S. Fustukian, and K. Buse. 2003. An introduction to global health policy. In *Health Policy in a Globalising World*, eds. S.K. Lee, K. Buse, and S. Fustukian, 3-17. Cambridge: Cambridge University Press.

Lehman B. 2002. *Making the world safe for biotech patents*. Boston: International Intellectual Property Institute.

Malby, S. 2002. Human dignity and human reproductive cloning. *Health and Human Rights* 6: 103-35.

Meldolesi, A. 2003. EU stalls on funding ES Cell Research. *Nature Biotechnology* 21: 588-9.

Monachello, J. 2003. The cloning for biomedical research debate: Do the promises of biomedical advances outweigh the ethical concerns? *Tulsa Journal of Comparative and International Law* 10: 591-623.

National Bioethics Advisory Commission. 1997. *Cloning Human Beings*. Available at: <http://www. georgetown.edu/research/nrcbl/nbac/pubs/cloning1/cloning.pdf> (Last accessed: 24 January 2005).
Nature Biotechnology. 2000. Human genome bombshell 18: 365.
Nature. 2002. Editorial. 1331.
Nuffield Council on Bioethics. 2002. *The ethics of patenting DNA*. London: Nuffield Council on Bioethics.
OECD. 2001. *Biotechnology statistics in OECD member countries* (September 13). Available at: <http://www.olis.oecd.org/olis/2001doc.nsf/c5ce8ffa41835d64c125685d005300b0/c1256985004c66e 3c1256ac600350f21/$FILE/JT00112476.PDF> (Last accessed: 24 January 2005).
Ontario Report to Premiers. 2002. *Genetics, testing and gene patenting: Charting new territory in healthcare*. Available at: <http://www.health.gov.on.ca:80/english/public/pub/ministry_reports/ geneticsrep02/report_e.pdf> (Last accessed: 24 January 2005).
Oriola, T. 2002. Ethical and legal issues in Singapore biomedical research. *Pacific Rim Law and Policy Journal* 11: 497-529.
Pattinson, S. 2002. Reproductive cloning: Can cloning harm the clone? *Medical Law Review* 10: 295-307.
Pattinson, S., and T. Caulfield. 2004. Variations and voids: The regulation of human cloning around the world. *BMC Medical Ethics* 5: 1-8.
Pollara, E. 2002. *Public opinion research into biotechnology: Sixth wave*. Presented to the Biotechnology Assistant Deputy Minister Coordinating Committee (BACC), Government of Canada.
Regnier, M.H., and B.M. Knoppers. 2002. International initiatives. *Health Law Review* 11: 67-71.
Resnik, D.B. 2001. DNA patents and human dignity. *Journal of Law, Medicine and Ethics* 29: 152-65.
Reuters. 2004. U.S.-led push for broad U.N. cloning ban crumbles. 19 November.
Saad, L. 2002. Cloning humans is a turn off to most Americans. *Gallup News Service*, 16 May. Available at: <http://www.genetics-and-society.org/analysis/opinion/detailed.html> (Last accessed: 18 April 2005).
Sarma, L. 1999. Biopiracy: Twentieth-century imperialism in the form of international agreements. *Temple International and Comparative Law Journal* 13: 107-36.
Scherer, F. 2002. The economics of human gene patents. *Academic Medicine* 77: 1348-67.
Scoop. 2004. ADHB challenges bio-tech co. over patent claim. 20 August. Available at: <http://www.scoop.co.nz/mason/stories/GE0408/S00076.htm> (Last accessed: 24 January 2005).
Sheremeta, L., and R. Gold. 2003. Creating a patent clearinghouse in Canada: A solution to problems of equity and access? Paper delivered at GELS Conference, February 6-8, Toronto, Ontario.
Stem Cell Week. 2002. Editorial, 11 & 18 March.
Straus, J. 1998. Bargaining around the TRIPS Agreement: The case for ongoing public-private initiatives to facilitate worldwide intellectual property transactions. *Duke Journal of Comparative and International Law* 9: 91-108.
Taylor, A. 1999. Globalization and biotechnology: UNESCO and an international strategy to advance human rights and public health. *American Journal of Law and Medicine* 4: 479-542.
The President's Council on Bioethics. 2002. *Human cloning and human dignity: An ethical inquiry* (Washington DC: PCB). Available at: <http://www.bioethics.gov/cloningreport/fullreport.html> (Last accessed: 24 January 2005).
Thomas, S., M. Hopkins, and M. Brady. 2002. Shares in the human genome: The future of patenting DNA. *Nature Biotechnology* 20: 1185-8.
Thurow, L. 2000. Globalization: The product of a knowledge based economy. *Annals of the Academy of Political and Social Science* 570: 19-31.
Traubman,T. 2002. Leaps of faith. *Tishrei/Israeli Times*, 16 September.
TRIPS Council considers public health, biodiversity. 2004. *Bridges Weekly Trade News Digest* 8(42). Available at: <http://www.ictsd.org/weekly/04-12-08/story1.htm> (Last accessed: 5 May 2005).
United Kingdom Royal Society. 2003. *Keeping science open: The effects of intellectual property policy on the conduct of science*. London: Royal Society.
UNEP (United Nations Environment Programme). 2002. Integrating environment and development: 1972-2002. Available at: <http://www.unep.org/GEO/geo3/pdfs/Chapter1.pdf> (Last accessed: 25 January 2005).

Walsh, V. 2002. Biotechnology and the UK 2000-05: Globalization and innovation. *New Genetics and Society* 21: 149-76.
Wertz, D., M. Regnier, and B.M. Knoppers. 2003. Stem cells in a pluralistic society: Consequences of proposed Canadian legislation. *GenEdit* 1:1. Available at <http://www.humgen.umontreal.ca/int/GE_Arch_v.cfm?an=2003&no=1> (Last accessed: 24 January 2005).
Wright, T.G. 2000. Second thoughts: How human cloning can promote human dignity. *Valparaiso University Law Review* 35: 1-35.

CHAPTER TEN

KERRY PETERSEN

THE RIGHTS OF DONOR-CONCEIVED CHILDREN TO KNOW THE IDENTITY OF THEIR DONOR

The problem of the known unknowns and the unknown unknowns

There is no consistent regulatory pattern controlling assisted reproductive technologies (ART) at the international or domestic levels and it remains an area of medicine steeped in ethical and moral dilemmas. At this point, the models of regulation range from ART statutes which provide comprehensive legislative schemes, ad hoc statutes which address specific issues, such as surrogacy laws and professional guidelines, to combinations of these models and zero legal regulation (Lee and Morgan 2001, 265-305). Until recently, ART children were not permitted access to information about donors, however, concerns about human rights issues and the personal implications of denying ART children with this information, has provided an impetus for the introduction of disclosure laws and policies in an increasing number of countries.

Birthrights are generally not a problem where family members are related through marriage, biology or adoption. The formal process of recording information about identity and personal history commences with birth registration. Once born, a child is registered and a birth certificate is issued which records information about the child's personal history. When a child is born from donated gametes in Australia or Britain, the personal history recorded on the birth certificate is socially constructed. The birth certificate records the social parents, but not the identity of the donor(s) or that the conception was assisted by a donor. There is no legal requirement to inform the child about the means of conception. The legal fiction provides ART children with legal status, facilitates social acceptance of the ART family (Frith 2001) and regulates family formation (Haimes 2002, 444).

This chapter examines the culture of secrecy which has influenced laws and policies regulating donor insemination (DI) and the rights of donor-conceived children to identifying information about their donors, who in practice are usually sperm donors "since the nature of ovum and embryo donation means that the donor

151

B. Bennett & G.F. Tomossy (eds.), Globalization and Health: Challenges for Health Law and Bioethics,
151-167.
© 2006 *Springer. Printed in the Netherlands.*

is rarely anonymous" (Forder 2000, 256). The chapter draws substantially on developments in Australia and Britain (Chalmers 2002; Lee and Morgan 2001; Petersen 2002; Szoke 2002), however, developments in other countries are also explored. Australia has a complicated system of ART regulation operating in a federal system (NHMRC 2004; FSA 2005; Saunders 2003; Szoke 2003). In Britain, ART is regulated by the *Human Fertilisation and Embryology Act* 1990 (HFE) and the Human Fertilisation and Embryology Authority (HFEA) which administers the HFE through a Code of Practice and a licensing system. In Victoria, the Infertility Treatment Authority (the ITA) plays a similar role in administering the *Infertility Treatment Act* 1995 to the HFEA but is a much smaller operation.

In this chapter it is shown that courts and parliaments in a number of jurisdictions have re-visited the assumption that donor anonymity is in the best interests of ART offspring and are granting them varying degrees of access to information about their donors. This gradual and incremental change is reinforced by a human rights discourse which prioritizes a child's right to a personal history and a heritage over the privacy claims of the social parents and the donor. Nevertheless, even when courts and parliaments permit ART children to have access to this information, most social parents do not tell their children about the means of their conception. As Donald Rumsfeld has famously stated "we know there are some things we do not know. But there are also unknown unknowns – the one's we don't know we don't know."[1] Herein lies the Catch 22: how can a person exercise the right to know their biological origins if they don't know that they don't know?

It is contended here, that the full implications of these new laws and policies for families will not be fully realized until social parents are confident that giving their children the opportunity to discover their true heritage will be more beneficial than harmful. However, even though the law could become pro-active and require the donor's name to be included on the birth certificate or impose a statutory duty on the social parents to tell their children about their conception, this proposal has not been a significant feature of the public debate.[2]

MEDICAL BACKGROUND

Before examining recent developments governing disclosure and non-disclosure, it is important to consider the medical background in which donor anonymity became the norm and later an assumption (Bennett 1997, 129).

Artificial insemination with donor sperm (DI), was commonly used in medical practice by Australian and British doctors in the late 1930s – that is, around the same time that eugenics and birth control were controversial issues. In those early years, DI was shrouded in shame and practised secretly. Fresh sperm was the only means of insemination available and the "indelicacy" of being so proximate to a stranger in these circumstances may have been a further reason for secrecy. The practice of DI was rejected and discouraged by many in part because of religious concerns and fears of eugenic implications. In addition, sperm donors were treated with some suspicion and the shame of male infertility contributed to the stigma attached to donor conception (Bennett 1997, 131-6; Frith 2001, 819; Jackson 2001, 212).

Little attention was paid to recording information about donors and if clinics kept records they generally destroyed them after five or seven years (Frith 2001, 820). It was standard for medical practitioners to advise couples to remain silent about the means of conception and thus concealment became entrenched in medical practice. This approach was adopted to protect the donor, the social parents and the child from social and legal consequences and to conceal the donation (Bennett 1997; Jackson 2001, 212; Klock, Jacob and Maier 1994). Doctors were concerned to protect themselves from allegations of unprofessional or even unlawful conduct. Furthermore, the social parents often committed an offence by entering false information on the birth certificate and registering the husband as the legal father.

Children born from donor conception were classified by law as illegitimate or *nullius filius* (son of nobody) and this lack of status raised the following questions: first, was the donor legally liable as the biological father for child support and could the child make a claim against the donor's estate? Second, what could the social parents do if the donor wanted to make parental claims on the child? Eventually, these problems were addressed by status of children legislation which predated ART statutes in Australia and the UK. The legislation creates a rebuttable presumption that the birth mother and her consenting partner are the legal parent(s) and extinguishes the legal tie with the donor.[3]

By the early 1980s anonymity was being questioned but still practised routinely. However, after adoption policies changed from secrecy to openness in the 1970s and 1980s, the tide also began to turn for children created through DI (Jackson 2001, 215). At the time, the debate on anonymity was conducted mostly within professional circles between "health professionals, who favored anonymity, and social work professionals, who favored openness" (Haimes 2002, 444). Nevertheless, most jurisdictions continued to retain non-disclosure laws and policies (Frith 2001).

TELLING: CONCEALMENT AND OPENNESS

The debate has continued and the technologies have become more and more sophisticated and there is still no definitive evidence demonstrating whether or not traditional anonymity practices are harmful to ART offspring (Shenfield and Steele 1997). However, increasingly social parents are more open with their children about the means of their conception (Daniels and Thorn 2001, 1792). The two sides of the debate can be summarized briefly as follows. Arguments in favor of providing a child with access to identifying information about biological parentage are based on the claim that access is in a child's best interests and this entitlement should be prioritized over the interests of all other parties. Arguments on the other side are based on the claim that access may not always be in the best interests of a child; and in any case, there are other people including the extended family who have equally important privacy claims and interests.

Although there is a greater social and legal recognition of a right to know one's origins (Steiner 2003; Jackson 2001; McWhinnie 2001, 425; Tarrant 2002, 338; Rumball and Adair 1999), the culture of secrecy means that a legal right to identifying information *per se* will continue to have limited value. Before considering reasons underlying this secretiveness, it is worth noting some of the implications which flow from terminology used in much of the debate. Klock, Jacob and Maier (1994, 482) argue that key words used to discuss this process of conception have been *secrecy* and *privacy* rather than *openness* and *disclosure* and that secrecy implies guilt or shame and openness implies honesty and frankness. Therefore, when we use the word secrecy we tend to assume there is something to hide, perhaps, in this case, infertility or sexual dysfunction; whereas openness has an entirely different connotation.

Taking this notion of secrecy further, the following scenario from an article by a leading scholar in the area, Erica Haimes, reveals a "behind the scenes" snapshot of DI which speaks volumes. Broadly, she portrays the biological *dramatis personae* in the following way:

- A woman who plans to conceive a child using the semen of someone other than her husband or partner.
- A man who, even from the preconception stage, plans to raise a child to whom he is not genetically related.
- Another man who plans to have a child with a woman he does not even know.   (Haimes 2002, 446)

Considered this way, DI may seem confronting, even bizarre, but in reality this representation is not unusual and it throws some light on why various people continue to be ambivalent and therefore secretive about DI.

Lee and Morgan (2001, 232) refer to the practice of "not telling" and observe that:

> The decision of what to tell children born of assisted conception has long stood as one of the most problematic aspects of technological creation. The balance between preserving the identity of the donor and fracturing the identity of the resulting child has produced one of the deepest of philosophical and pragmatic tensions.

Some still may not tell their children about the donor insemination to protect the child from the shame of infertility. As Bennett (1997, 129) says: "Concerns over stigmatizing the DI child and family and the emotional difficulties of publicly acknowledging male infertility have all contributed to providing incentives for secrecy in DI." However in a study of Dutch families, Brewaeys et al. found that "very few parents mentioned the stigma associated with male infertility as an important motive." Instead, it seems that "uncertainties about the use of donor gametes rather than the taboo surrounding male infertility is the DI couples' major motive in their choice for secrecy" (Brewaeys et al. 1997a, 1595). This supports the view that ambivalence about using DI is another important reason for concealment practices and policies.

Notably, most lesbian couples and single women who use donated sperm are more likely to tell a child that a donor has been used than heterosexual couples. As these families usually live in alternative family structures with no male parent, the mothers have little choice but to be open about the means of conception including the use of a donor (Brewaeys et al. 1997b, 1351). It also seems that the non-biological father may be more reluctant to reveal the donation to their children than the biological mother (Brewaeys et al. 1997a, 1596). Likewise, in lesbian families, the non-biological mother was more often reluctant to gain information about the donor. Brewaeys et al. suggest that the non-biological parent in both homosexual and heterosexual families may feel more vulnerable, and may experience the donor as a threat to their position in the family and to their relationship with the child (ibid.).

Other factors which help to explain parental reluctance to tell children include fears the child will reject their father because of the perceived stigma attached to using assisted conception and because of the lack of a genetic relationship (Tarrant 2002, 338). Parents also expressed concerns that the child might be rejected by others for being different. As well, parents have justified "not telling" to protect the donor from unwanted disruption, contact or responsibility and to maintain family privacy and stability (Vanfraussen, Ponjaert-Kristoffsen and Brewaeys 2001, 2020-3; Tarrant 2002, 338). Even when not averse to their child knowing, parents are unsure about how and when to tell the child about their conception (Vanfraussen Ponjaert-Kristoffsen and Brewaeys 2001, 2019; Salter-Ling, Hunter and Glover 2001). However, even though most social parents do not tell their children they were conceived by DI, they (particularly the mother) will often tell another person such as a family member or close friend. Generally someone other than the social parents will know the family secret and this revelation can be damaging for a person to receive later in life (Cook 2002; Klock, Jacob and Maier 1994; Leiblum and Aviv 1997). In sum, social parents seem confused and not particularly well informed on the issue of "telling or not telling."

Nevertheless, according to Tarrant, attitudes are changing with a "growing rejection of the shame that motivates or results from the silence around assisted conception and a growing claim that secrecy is in the interests of the adults involved and not the child" (Tarrant 2002, 338). The social parents may be the ones who are best placed to decide how their family should address this difficult question (Sheffield and Steele 1997, 392); however, with increasing legal acceptance of the right to know one's origins, inevitably more pressure will be placed on parents to endorse the value of truthfulness.

INTERNATIONAL TRENDS[4]

Human rights developments

Contemporary developments in human rights law are based on the right to know one's origins, the right to privacy and protection of the public interest. The United Nations *Convention on the Rights of the Child* (CROC) and the *European Convention of Human Rights* (ECHR) provide a foundation for arguments supporting the right to know one's origins and the increasing acceptance by courts of this right imposes a higher onus on the custodians of this information to justify secrecy.[5] Articles 7 and 8 of the CROC, the blueprint for children's rights, are particularly relevant as two of its provisions concern issues of identity. They state that:

> 7 (1) The child shall be registered immediately after birth and shall have the right from birth to a name, the right to acquire a nationality and, as far as possible, the right to know and be cared for by his or her parents.
>
> (2) States Parties shall ensure the implementation of these rights in accordance with their national law and their obligation under the relevant international instruments in this field, in particular where the child would be otherwise stateless.
>
> 8 (1) States Parties undertake to respect the right of the child to preserve his or her identity, including nationality, name and family relations as recognized by the law without unlawful interference.
>
> (2) Where a child is illegally deprived of some or all of the elements of his or her identity, States Parties shall provide appropriate assistance and protection, with a view to speedily re-establishing his or her identity.

Australia became a party to the CROC after signing the instrument in August 1990 and ratifying it in December 1990. It is well established that the provisions of an international treaty do not form part of Australian law unless specifically incorporated into domestic law and therefore, the CROC is not directly applicable. However, as this treaty has been ratified, there is a legitimate expectation that officials will comply with it.[6] The full implications of CROC's relevance to Australian ART children in states where there is no statutory right to identifying information is not yet clear (Jones and Basser Marks 2001).[7]

The right to respect for privacy and family life contained in Article 8 of the ECHR has been highly instrumental in the recognition of a right to know one's origins in European countries and in Britain where the ECHR applies courtesy of the *Human Rights Act* 1998 (UK). Article 8 provides that:

> (1) Everyone has the right to respect for his private and family life, his home and his correspondence.
>
> (2) There shall be no interference by a public authority with the exercise of this right except such as is in accordance with the law and is necessary in a democratic society in the interests of national security, public safety or the economic well-being of the country, for the prevention of disorder or crime, for the protection of health or morals, or for the protection of the rights and freedoms of others.

Initially, Article 8 was used to assist people who had been in care and were seeking information about their biological parent(s). As the word "everyone" is included in Article 8(1) it is important to note that both parents and children have a right to respect for private and family life. However, under Article 8(2), a public authority may interfere with these rights when important public interests are at stake.

In *Gaskin v. United Kingdom*[8] the applicant, relying on Article 8 of the ECHR, successfully applied to the European Court of Human Rights for access to the records of his time spent in foster care to help him overcome his problems and learn about his past. The Court held that the procedures followed by the local authority failed to secure respect for Gaskin's private and family life as required by Article 8 and his interest in learning about his origins outweighed the public interest in managing an efficient child care system. The Court took the view that a person in Gaskin's situation has a "vital interest, protected by the Convention, in receiving the information necessary to know and to understand their childhood and early development."[9] This interpretation of Article 8 was approved by the European Court of Human Rights in *Mikulic v. Croatia*[10] where the applicant who was born "out of wedlock," filed a civil suit against the respondent on the basis that the delays in establishing paternity left her uncertain about her personal identity. The Court found that Article 8(1) can sometimes embrace aspects of a person's physical and social identity and that in this case there was a direct link between establishing paternity and the applicant's private life. Furthermore, the interest of the applicant in discovering an *important* aspect of personal identity was more compelling than the interest of a third party in refusing to be compelled to submit to paternity testing. However, the European Court of Human Rights distinguished *Gaskin* and *Mikulic* from *Odièvre v. France*[11] because the mother in *Odièvre v. France* had abandoned the applicant at birth and expressly requested that information about the birth remain confidential under the French system of anonymous births.[12] In *Odièvre v. France* the Court found that Article 8 protected a right to identity, but it did not override the competing privacy rights of a mother in these circumstances particularly when the state has a public health interest in assisting pregnant women to deliver children safely.

Parallel developments concerning identity rights in the Dutch and UK courts have preceded legislative changes in these two jurisdictions.

The Netherlands

In the 1990s, the Dutch courts recognized a right to know one's origins in the *Valkenhorst* cases. These cases concerned people born to "unwed" women and placed in care in the 1920s who sought to discover the identity of their fathers from files kept by the Catholic charitable institution Moederheil (Forder 1996, 357). In the first case, *Roovers v. Hertogenbosch,*[13] the Gerechtshof's-Hertogenbosch, a court of second instance, held that the applicant, Maria Roovers, whose mother died in 1944, had a right to know the identity of her father because this consideration weighed more heavily than her deceased mother's right to privacy. After this case, the respondent institution Valkenhorst took over the role of the Moederheil and changed its policy to permit disclosure of the mother's identity in all circumstances. The Valkenhorst was also willing to disclose the father's identity unless a live mother refused to consent. The new policy was challenged in the second Valkenhorst case *Monteyne/de Ruyter/Derks v. Valkenhorst*[14] where the Dutch Supreme Court ruled that a qualified right to know one's origins was part of Dutch law, and that basic rights, such as respect for private life, constitute part of the implicit freedom of personality which includes the right to know the identity of one's parents. However, the Court also ruled that it is a qualified right which must be weighed against other rights such as parental privacy interests and "social function" interests (ibid.). In addition, the Court acknowledged the relevance of Article 7 of the CROC.

United Kingdom

Anonymity laws were reviewed recently in the UK and a public consultation document was issued in December 2001 (UK, Department of Health 2001). The consultation ran until July 2002. While the consultation was taking place, *R (on the application of Rose and another) v. Secretary of State for Health and the Human Fertilisation and Embryology Authority*[15] was heard by Scott Baker J in the High Court. The applicants, Ms Rose, born in 1972, and EM born in 1996 (represented by her mother as litigation friend), who were both conceived with donated sperm, sought judicial review of the Secretary of State's decision to refuse to grant them access to information about their donors, or the establishment of a contact register, because the Department of Health consultation exercise was still in process at that time. The claimants argued that the provisions of the ECHR guaranteed them a right of respect for private and family life and non-discrimination. However, at the case management conference, Scott Baker J decided to confine the issues to be tried because of the consultation exercise that was taking place at the time.[16] As far as Article 8 of the ECHR was concerned, he decided that the only question for the Court at that time was to consider whether or not it was engaged and, if so, he would adjourn the hearing in order to proceed with the balancing exercise under Article 8(2) and decide if there had been a breach.[17] In any event, the Court decided that Article 8 was engaged both with regard to identifying and non-identifying information but the question a breach was not addressed.[18] Scott Baker J said, "The bottom line…is that the donor provided half of each Claimant's genetic identity and it is this that creates the interest of the Claimant to seek information about him."[19] The results of Department of Health consultation were not public when the *Rose* judgment was handed down on 26 July 2002. However, as will be seen the *Rose* case will not need to be resumed because of subsequent legislative developments discussed below.

LEGISLATIVE DEVELOPMENT

Proposals for changing anonymity practices have raised concerns about a shortage of gametes and consequent reduction of ART services from accredited clinics. Reference is often made to the decrease in semen donations in Sweden after the passage of a 1984 Swedish law[20] which permitted the donor's identity to be disclosed. However, it is claimed that this was corrected later with the recruitment of a more mature group of donors (Daniels and Lalos 1995, 1875).

In 1996, the Victorian medical profession was worried about the effect disclosure laws could have on the willingness of men to donate sperm and there was a slight decline in donations after the 1995 legislation took effect in Victoria. However the cause of the decline is not clearly related to the disclosure provision in the Act (Infertility Treatment Authority 2000, 17). British medical practitioners have expressed similar reservations. At the 2002 British Medical Association's Conference, the members voted narrowly against removing the anonymity provision[21] because of the apprehension that people would be deterred from making donations to clinics if they could be identified and the consequences this could have for fertility clinics and people seeking treatment.[22] In January 2005, the UK Department of Health announced that it was launching a national campaign to recruit egg and sperm donors because of the shortage of donors, which has been exacerbated by the new regulations coming into force in April 2005 (Progress Educational Trust 2005, 290).

When the British government announced in January 2003 that as a first step it was considering permitting children conceived by donation since 1990 to request non-identifying information about the donor, the Health Minister, Hazel Blears said: "We are especially concerned about the possible effect on donor numbers of removal of anonymity...and thus on the number of people able to receive treatment."[23]

In the Netherlands, a government-commissioned survey conducted in the 1990s found that 50 percent of current sperm donors said that once the Dutch *Storage and Disclosure of Information on Gamete Donors Act* 2001 came into force they would not make any further donations. It was also found that the proposed legislation was poorly understood in the general community and the most common reason given for being a donor was "to help someone else have a child" (Forder and Sumner 2003, 280).

Internationally, regulation ranges from those laws which prohibit disclosure, to those which provide a qualified or an absolute right to identifying information about the donor. However, the content and nature of the disclosure provisions vary considerably from jurisdiction to jurisdiction. The following models illustrate some of these approaches.

Disclosure prohibited

In most European countries donor anonymity is the norm. Denmark,[24] France, and Norway are examples of this model.

Qualified right to identifying information

Countries such as the Netherlands and Canada have given ART offspring a qualified right to identifying information.

The Dutch Parliament tabled the *Storage and Disclosure of Information on Gamete Donors Bill* in 1993 and finally passed the Act on 25 May 2001. The disclosure provisions become operative in 2004 (Forder and Summer 2003; Progress Educational Trust 2004, 261).[25] The *Storage and Disclosure of Information on*

Gamete Donors Act 2001 requires that fertility clinics store relevant information for eighty years; not accept anonymously donated sperm samples; and they must give information about donors to the Gametes Donor Foundation (the Foundation) if the donation results in a pregnancy.

The Act establishes procedures for cases where the donor does not want to be identified. The following three categories of information can be made available to different parties, irrespective of the biological mother's wishes:

1. *Medical information* – this is to be made available upon application by the child's doctor – irrespective of age.

2. *Non-identifying information* – this is to made be available upon application by a child of twelve years or more who knows or suspects having been conceived by DI or egg donation. If the child is under sixteen years, the parents must be notified that the information has been disclosed, and the child told that the parents were notified. When a child is under twelve years parents may be supplied with the information.

3. *Identifying information about the donor* – this is to be made available to a child who has reached at least sixteen years, if the donor gives written permission. However, if the donor withholds permission or makes no response the information may be supplied after the interests of the child and the donor have been considered. If the clinic intends to provide identifying information, the donor must be notified and can lodge an objection with the Foundation. If the Foundation decides to disclose the information, the donor may appeal to the courts. Similar processes are available to relatives where the donor is deceased or untraceable. In these cases, the interests of the deceased or missing donor *only* are taken into account in the decision making process.

Dutch clinics have offered a double-track system for some years (Pennings 1997). Under this system, *prospective parents* seeking donor sperm may go to Counter A or Counter B at the clinic, and *sperm donors* may offer to provide sperm for Counter A or Counter B applicants. Counter A provides an anonymous service and Counter B provides an open service. It seems that Counter B services are mostly used by lesbian and single women where information is available to each party as they are generally open about their child's biological origins (Forder 2000, 257). This system will continue to operate under the new legislative regime, however, Counter A donors will be subject to the disclosure processes in the new Act in cases where the information is sought.

The Canadian disclosure provision is limited in its application. Section 15(1) of the *Assisted Human Reproduction Act* 2004[26] prioritizes the privacy rights of the donor and permits clinics to provide an ART child with identifying information *only* with the donor's consent (Progress Educational Trust 2003, 232).

Absolute right to identifying information

These laws have been passed in Sweden, Victoria and Britain. In New South Wales, a consultation Bill is under consideration.

Under Swedish law, a "sufficiently mature" child conceived through insemination may be "apprised of the information [about] the sperm donor which is recorded in the hospital's special register."[27] Oocyte donation has also been permitted in Sweden since January 2003.

In Victoria, the *Infertility Treatment Act* 1995 provides people conceived with donor gametes an unqualified right to identifying information about their biological parentage once reaching eighteen years and after they have received counseling.[28] If they wish to access this information earlier the application must be made through their parents and the release of this information is still subject to the donor's consent.[29] The Act also gives donors the right to identifying and non-identifying information about a resulting child. If the ART child is under eighteen years the social parents/guardian *must* give consent before this information can be provided to the donor. Once the ART offspring is eighteen, he or she must consent before the information can be given to the donor. This poses the risk that the request for the consent will alert an adult person about the means of their conception later in life and it could be emotionally harmful. They also lose the option of not knowing they were conceived by DI.

Information can be obtained from earlier donors, with their consent, through a system of voluntary registration administered by the Victorian Infertility Treatment Authority (ITA). Victoria has recorded every birth resulting from a donor treatment procedure since 1988 and the ITA now maintains four registers including a voluntary register (Infertility Treatment Authority 2003, 14-15).[30] The full implications of disclosure rules have been unclear because of some issues arising from earlier regulations. The Victorian Law Reform Commission has declined to make a recommendation for retrospective access to identifying information. However, the ITA has advised clinics that from June 2006 they will no longer be permitted to use gametes donated prior to January 1998 unless a family who has already conceived through donor gametes wants to have more children using the same donor.[31]

New South Wales has also moved in this direction with the publication of a Consultation Draft Bill in December 2003.[32] A core objective of this Consultation Draft Bill is to give people born as a result of donor conception access to identifying and non-identifying information about their genetic parentage from the information which will be recorded on a central donor register to be administered by the Department of Health.[33] The proposed right will not be retrospective. Moreover, ART clinics will be obliged to inform the donors about the rights of ART offspring and other persons who are entitled to have access to information which is on the central donor register.[34]

Until recently, the provisions of the *Human Fertilisation and Embryology Act* 1990 (UK) only allowed the HFEA to give limited information about donors to ART offspring in two situations: at sixteen years, if a person wanted to know if they could be related to someone they intend to marry (s 31(6)); and at eighteen years, if a person wanted to know if they had been born as a result of infertility treatment using donated gametes or embryo (id., s 31(3)). The HFEA maintains a confidential register of all donors and all people receiving treatment at licensed clinics (id., s 31(1)). These information rights were extended when the *Human Fertilisation and Embryology Authority (Disclosure of Donor Information) Regulations* 2004 removed anonymity for gametes and embryo donors and became effective in 2005. They provide that a person born from ART treatment, who has reached eighteen years, may apply to the HFEA for access to non-identifying (Reg 2(2)) and identifying information (Reg 2(3)) from the HFEA register. Although the right is not retrospective, people who have made donations in the past may agree to identifying information being made available by giving their details to clinics after March 2005 (Reg 2(3)). In addition, a voluntary scheme has been established for children born prior to 1 August 1991 when data began to be collected for the HFEA register.[35] This legislation prioritizes the child's right to information about his or her genetic origins over a donor's right to privacy (Progress Educational Trust 2004, 242).

CONCLUSIONS

The assumption that donor anonymity should be the cornerstone of infertility programs developed within the privileged doctor-patient relationship and was justified on the grounds that it provided protection for both the donor and social parents and was in the best interests of ART offspring. Even after legislation in Australia and Britain dealt with concerns about legal status, parentage and other potential suits, secrecy about the donor's identity was maintained officially through laws and policies and unofficially by social parents.

In the public debate, the focus has been on the privacy of the donor (and the donor's family) and the best interests of ART children. In addition, practical and ethical concerns about the effect of abolishing anonymity rules on donor recruitment and infertility services have acted as a brake on change in some jurisdictions.

There is no compelling evidence one way or the other as to how children are affected by knowing or not knowing that they were conceived by DI and the results of longitudinal research on this question will take time. However, the experience of adoptees and people placed in state homes, has been with us longer than the experience of ART offspring, and the *Rose* decision suggests that the need for a name, a personal identity, a personal history and heritage is profound. At the same time, it must be said that non-disclosure laws and strategies served a number of important functions before in vitro fertilization (IVF) by donor became as commonplace as it is today. They have helped to camouflage societal and personal ambivalence about infertility, uncertainty about using donated gametes and uncertainty about conceiving a child through artificial means. They have also played an important role in regulating family formation by giving social parents the opportunity to mimic the dominant heterosexual two-parent family model

(Charlesworth 1993, 63-64). Heterosexual two-parent families can pass as "normal" biological families fairly easily – particularly if they can conceal the role of the biological stranger in the creation of their child. Lesbian and single parent families, on the other hand, cannot pass as "normal" biological families and have little to gain by trying to conceal the means of conception. The change of emphasis from adults rights to children's rights suggests that the underlying purposes served by anonymity policies and laws, have diminished in importance, at least in some countries.

The legislative overview of Western jurisdictions in this chapter, demonstrates that there is a gradual shift towards laws permitting people to have access to identifying information about donors but, as is typical of ART regulation, there is no coherent pattern in these developments and the nature of the right varies greatly from country to country. It is fairly clear that the recognition of a right to a personal identity carved out in human rights law has been the trump card because the recognition of this right (albeit a qualified right) has placed pressure on courts and legislatures to address the issue. The human rights dimension is particularly significant because it elevates the discussion from a "shopping list" of pros and cons to an intellectual and legal debate.

A deep analysis of the reasons for hiding the role of the biological stranger in creating a family and for the state to collaborate in this masquerade goes beyond the remit of this chapter, but it is clear from the discussion on the culture of secrecy that these reasons are extremely complex and have been a driving political force. As Haimes says, donor anonymity policies are clearly "about more than the individual's 'identity crisis' than it is commonly portrayed as being" (Haimes 2002, 447). For human rights charters and statutes to be truly effective they will need to be supported by appropriate administrative structures and support. It is likely, therefore, that the problem of the "known unknowns and the unknown unknowns" will continue until DI has more personal and social legitimacy (Cook 2002, 222; Haimes 2002, 447). Nevertheless, statutes and judicial decisions have functional as well as symbolic value and the human rights developments that have already taken place in the courts have raised consciousness about the importance of the issue. They also impose a responsibility on politicians, judges and the community to acknowledge that people created with assistance of state approved infertility programs have rights, albeit qualified or absolute rights, to information about their biological origins.

NOTES

I would like to thank Tessa Hanscombe for her invaluable help with the research for this chapter; Professor Caroline Forder from the University of Maastricht, the Netherlands for kindly helping me with the section on Dutch law; Professor Martin Richards from the Centre for Family Research, Cambridge UK, for his comments on an earlier draft and Dennis Warren, Law Librarian, La Trobe University, for his support and patience in assisting with library technologies. An earlier version of this chapter was presented at the 28th International Congress on Law and Mental Health in Sydney, Australia (2003).

1. This quote draws on comments made by US Defence Secretary Donald Rumsfeld at a news conference in February 2002. The full quote, which won the annual British Plain English Campaign in 2003, can be accessed at <http://www.plainenglish.co.uk/pressarchive.html#Anchor-Donal-1943> (Last accessed: 31 March 2005).

2. Two decades ago, Dame Mary Warnock proposed that the words "by donation" should be on entered on the birth certificate. See Warnock (1984, paras 4.25, 6.8). This proposal was defeated in the House of Commons.

3. *Family Law Act* 1975 (Cth), s 60H; *Status of Children Act* 1996 (NSW), s 14; *Status of Children Act* 1974 (Vic), ss 10A-F; *Status of Children Act* 1978 (Qld), ss 13-18; as amended by No 64 of 1988; *Family Relationships Act* 1975 (SA), ss 10a–10e as amended by No 102 of 1984 and No 2 of 1988; *Artificial Conception Act* 1985 (WA), ss 5-7; *Status of Children Act* 1974 (Tas), ss 10a–10c as amended by No 122 of 1985; *Status of Children Amendment Act* 1985 (NT), ss 5A-5F; *Artificial Conception Ordinance* 1985 (ACT) ss 5-6; *Family Reform Act* 1987 (UK) ss 27-28.

4. The regulatory system adopted in the United States is not included in the chapter. There is no federal uniform legislation in the US – various state acts and guidelines from the medical profession constitute the major form of regulation. Some private clinics provide information about the identity of the donors.

5. See: United Nations. *Convention on the Rights of the Child*. 1990. Australian Treaty Series, No. 4. Available at: <http://www.austlii.edu.au/cgi-bin/disp.pl/au/other/dfat/treaties/1991/4.html?query= child>. (Last accessed: 2 May 2005). Council of Europe. *Convention for the Protection of Human Rights and Fundamental Freedoms*. 1950. European Treaty Series, No. 5. Available at: <http://www.echr.coe.int/Convention/webConvenENG.pdf> (Last accessed: 2 May 2005).

6. *Minister for Immigration and Ethnic Affairs v. Ah Hin Teoh* (1995) 183 CLR 273 at 287 per Mason CJ and Deane J. This principle is still binding in Australia even though members of the Hugh Court have suggested that the 'legitimate expectation' principle may be reconsidered in the future (*Re Minister for Immigration and Multicultural Affairs: Ex parte Lam* (2003) 214 CLR 1).

7. The South Australian Council on Reproductive Technology (SACRT) supported donors having the right to access identifying information about their gametes donor because of Article 8 of the CROC (South Australia, 2000).

8. [1990] FLR 167. Gaskin's case was heard by the European Court of Human Rights as it predated the *Human Rights Act* 1998 (UK).

9. [1990] FLR 167 at para 49.

10. [2002] 1 FCR 720.

11. [2003] 1 FCR 621.

12. The mother signed a form at the Health and Social Security Department relinquishing the child for adoption and requesting that the birth be kept secret.

13. Gerechtshof's-Hertogenbosch. 18 September 1991. Nederlands Jurisprudentie 1991, 796. Cited in Forder (1996).

14. Dutch Supreme Court. 15 April 1994. Nederlands Jurisprudentie 1994, 608. Cited in Forder 1996.

15. [2002] EWHC 1593.

16. Id., para 16 per Scott Baker J.

17. Id., para 20 per Scott Baker J.

18. Id., para 61 per Scott Baker J.

19. Id., para 38 per Scott Baker J.

20. 1984 Swedish Law No.1140.

21. Available at: <http://www.bma.org.uk/ap.nsf/Content/__Hub+resolutions> Resolution 593: 04.07.0 (Last accessed: 5 September 2004).

22. One media report claimed infertility clinics were are reporting a severe shortage of egg and sperm donation in September 2004, before the new laws had come into effect. See: <http//www.channel4./com/news/2004/09/week_2/8_donor.html> (Last accessed: 20 September 2004).

23. See: <http://www.dh.gov.uk/PublicationsAndStatistics/PressReleases/PressReleasesNotices/fs/en?CONTENT_ID=4024444&chk=LlIBuN> (Last accessed: 31 July 2005).

24. The Danish Council of Ethics advocated the abolition of anonymity for gametes donors on the basis that the right to know one's origins comes within the protection of the right to self determination. See, Council of Ethics 1995, 67.
25. Act of 25th May 2002, *Staatsblad 2002*, nr.240, Article 14. Cited in Forder and Sumner (2003); Progress Educational Trust (2004).
26. *Assisted Human Reproduction Act* SC 2004, c.2; at the time of writing the regulations had not been passed. As the federal government does not have jurisdiction over maintenance claims, donors would possibly be liable for maintenance until the provinces take some action to give them immunity. More details will presumably be inserted into the forthcoming regulations.
27. 1984 Swedish Law No. 1140.
28. *Infertility Treatment Act* 1995 (Vic) ss 78, 79.The disclosure rules in Western Australia were also changed by amendment in 2004. See *Human Reproductive Act* 1991 (WA) s.49(2)(a)-(e).
29. Ibid., ss 74, 75(2).
30. The four registers include information about donor procedures relating to egg, sperm and embryo donation. They are: the 1984 Central Register, the post-1988 Donor Treatment Procedure Register, the pre-1988 Donor Treatment Procedure Information Register and the Voluntary Register.
31. For a recent examination of these Victorian provisions see VLRC (2005, ch. 5).
32. *Assisted Reproductive Technology Bill* 2003 (NSW). The Consultative Draft Bill includes the following provisions:

> *Access by donor offspring who has reached 18 years*: These offspring are entitled to receive identifying and non-identifying information about the donor recorded on the central donor register. Furthermore, they may receive non-identifying information regarding other offspring of that donor *and* any identifying information about the other offspring where there is a registered consent (cll 38(3), 40(2)).

> *Access by parents or guardian where ART offspring under 18*: Non-identifying information about the donor and other offspring of the donor may be given to parents or guardians, but identifying information may only be disclosed in urgent circumstances and where this information would be otherwise unavailable (cl 39).

> *Access by donor to information about ART offspring*: Non-identifying may be given to the donor. Unlike Victoria, the NSW Director General may seek consent from the adult offspring but only if it necessary to promote welfare and best interests of one of the parties (cl 40).

33. Id., cll 34, 35.
34. Id., cl 14.
35. See <http://news.bbc.co.uk/1/hi/health/3642437.stm> (Last accessed: 2 May 2005).

REFERENCES

Bennett, B. 1997. Gamete donation, reproductive technology and the law. In: *Intersections: Women on law, medicine and technology*, ed. K. Petersen, 127-144. Aldershot: Ashgate.
Brewaeys, A., S. Golombok, N. Naaktgeboren, J.K. de Bruyn, and E.V. Van Hall. 1997a. Donor insemination: Dutch parents' opinions about confidentiality and donor anonymity and the emotional adjustment of their children. *Human Reproduction* 12(7): 1591-7.
Brewaeys, A., I. Ponjaert, E.V. Van Hall, and S. Golombok. 1997b. Donor insemination: Child development and family functioning in lesbian mother families. *Human Reproduction* 12: 1349-59.
Chalmers, D. 2002. Professional self-regulation and guidelines in assisted reproduction. *Journal of Law and Medicine* 9(4): 414-28.
Charlesworth, M. 1993. *Bioethics in a liberal society*. Melbourne: Cambridge University Press.
Cook, R. 2002. Villain, hero or masked stranger: Ambivalence in transactions with human gametes. In: *Body law and lores,* eds. A. Bainham, S. Day Sclater, and M.Richards, 211-27. Oxford: Hart Publishing.

Daniels, K., and O. Lalos. 1995. The Swedish Insemination Act and the availability of donors. *Human Reproduction* 10(7): 1871-4.

Daniels, K., and P. Thorn. 2001. Sharing information with donor insemination offspring: A child-conception versus a family-building approach. *Human Reproduction* 16(9): 1792-6.

Denmark. Council of Ethics. 1996. *Eighth Annual Report 1995.*

Forder, C. 1996. The Dutch report. In: *The International Survey of Family Law 1994 Edition*, ed A. Bainham, 357-60. The Hague: Martinus Nijhoff.

———. 2000. The Dutch report. In: *The International Survey of Family Law 2000 Edition*, ed A. Bainham, 239-61. Bristol: Family Law.

Forder, C., and I. Sumner. 2003. The Dutch report. In *The International Survey of Family Law 2003 Edition*, ed A. Bainham, 263-321. Bristol: Jordans.

Frith, L. 2001. Gamete donation and anonymity: The ethical and legal debate. *Human Reproduction* 16(5): 818-24.

FSA (Fertility Society of Australia). 2005. *Code of Practice for Assisted Reproductive Technology Units.*

Haimes, E. 2002. When transgression becomes transparent: Limiting family forms in assisted conception. *Journal of Law and Medicine* 9(4): 438-48.

Infertility Treatment Authority, Victoria. 2000. *Third Annual Report.*

———. 2003. *Sixth Annual Report.*

Jackson, E. 2001. *Regulating reproduction: law, technology and autonomy.* Oxford: Hart Publishing.

Jones, M., and L. Basser Marks, eds. 2001. *Children on the agenda: The rights of Australia's children.* Australia: Prospect Media.

Klock, S.C., M.C. Jacob, and D. Maier. 1994. A prospective study of donor insemination recipients: secrecy, privacy and disclosure. *Fertility and Sterility* 62(3): 477-84.

Leiblum, S.R., and A.L. Aviv. 1997. Disclosure issues and decisions of couples who conceived via donor insemination. *Journal of Psychosomatic Obstetrics and Gynacology* 18(4): 292-300.

Lee, R.G., and D. Morgan. 2001. *Human fertilisation and embryology: Regulating the reproductive revolution.* London: Blackstone Press.

McWhinnie, A. 2001. Gamete donation and anonymity: Should offspring from donated gametes continue to be denied knowledge of their origins and antecedents? *Human Reproduction* 16(5): 807-19.

NHMRC (National Health and Medical Research Council). 2004. *Ethical guidelines on the use of reproductive technology in clinical practice and research.* AGPS: Australia.

Pennings, G. 1997. The double track policy for donor anonymity. *Human Reproduction* 12(12): 2839-44.

Progress Educational Trust. *BioNews* 2003: 232. Available at: <http://www.BioNews.org.uk> (Last accessed: 3 November 2003).

———. *BioNews* 2004: 242. Available at: <http://www.BioNews.org.uk> (Last accessed: 28 Janurary 2004).

———. *BioNews* 2004: 261. Available at: <http://www.BioNews.org.uk> (Last accesssed: 7 June 2003).

Petersen, K. 2002. The regulation of assisted reproductive technology: A comparative study of permissive and prescriptive laws and policies. *Journal of Law and Medicine* 9(4): 483-97.

Rumball, A., and V. Adair. 1999. Telling the story: Parents' scripts for donor offspring. *Human Reproduction* 14(5): 1392-9.

Salter-Ling, N., M. Hunter, and L. Glover. 2001. Donor insemination: Exploring the experience of treatment and intention to tell. *Journal of Reproductive and Infant Psychology* 19(3): 175-87.

Saunders, D. 2003. The Australian experience of self-accreditation. In: *The regulation of assisted reproductive technology*, eds J. Gunning and H. Szoke 225-30. Aldershot: Ashgate.

Shenfield, F., and S.J. Steele. 1997. What are the effects of anonymity and secrecy on the welfare of the child in gamete donation? *Human Reproduction* 12(2): 392-5.

Steiner, E. 2003. *Odièvre v. France*: Desperately seeking mother – anonymous births in the European Court of Human Rights. *Child and Family Law Quarterly* 15(4): 425-48.

Szoke, H. 2002. The nanny state or responsible government? *Journal of Law and Medicine* 9(4): 470-82.

———. 2003. Australia: A federated structure of statutory regulation of ART. In: *The regulation of assisted reproductive technology*, eds. J. Gunning, and H. Szoke, 75-94. Aldershot: Ashgate.

Tarrant, S. 2002. State secrets: Access to information under the Human Reproductive Technology Act 1991 (WA). *Journal of Law and Medicine* 9(3): 336-46.

United Kingdom. Department of Health. 2001. *Donor information consultation: Providing information about gamete or embryo donors*. Available at: <http://www.doh.uk/gamete donors/document.htm> (Last accessed: 25 September 2002).

Vanfraussen, K., I. Ponjaert-Kristoffsen, and A. Brewaeys. 2001. An attempt to reconstruct children's donor concept: A comparison between children's and lesbian parents' attitudes towards donor anonymity. *Human Reproduction* 16(9): 2019-25.

VLRC (Victorian Law Reform Commission). 2005. *Assisted Reproductive Technology & Adoption. Position Paper Two: Parentage*. Melbourne: VLRC.

Warnock, M. 1984. *Report of the Commission of Inquiry into Human Fertilisation and Embryology. 1984*. Cmnd 9314. London: HMSO.

CHAPTER ELEVEN

JOHN HARRINGTON

GLOBALIZATION AND ENGLISH MEDICAL LAW

Strains and contradictions

In this chapter I wish to trace some of the consequences for English medical law[1] of the globalization of health care provision, especially as it concerns the UK's National Health Service. In the first half of the paper I offer an outline of what I mean by globalization in the context of health care. I then consider the manifestations of this process in two areas of medical practice: international trafficking in organs, and so-called health tourism within the European Union. In conclusion I discuss the generic stresses imposed upon the law by the uneven developments in the two areas considered. It can be seen that the tension between relatively recent global economic liberalization and the more traditional welfarist paternalism of the nation state is replicated in the changing caseload and sometimes incoherent doctrines of medical law.

WHAT IS GLOBALIZATION IN THE CONTEXT OF HEALTH CARE?

In response to the enthusiastic evocation of globalization by politicians and scholars during the 1990s (Giddens 1998), more recently commentators have questioned the extent and depth of the phenomenon (Henwood 2003, 148). They have also doubted the novelty of globalization, arguing that it is merely a return to pre-First World War patterns of trade (Petras 1999; Sutcliffe 2002, 52-4). The demise of the nation state, predicted by some, is also unlikely. The state and its law are vital to globalization: guaranteeing a compliant labor force and a benign fiscal regime for inward investors, as well as opening up new opportunities for profit through the privatization of public assets and the protection of intellectual property (Wood 2002).

Caution is well advised, therefore, in charting the effect of globalization on health care provision and on medical law in particular. Nonetheless a number of contemporary trends can be accommodated under a loose, descriptive concept of globalization: the transnationalization of production; growing free movement of

B. Bennett & G.F. Tomossy (eds.), Globalization and Health: Challenges for Health Law and Bioethics,
169-185.
© 2006 *Springer. Printed in the Netherlands.*

consumers, if not of workers; commodification of the human body and of formerly state-funded health care; persisting inequality between core and periphery in the world economy; uneven normative convergence with a predominance of economic law over other branches at international level. Given that the contemporary era is preeminently one of capitalist globalization I adopt a consistent perspective on these phenomena rooted in theories of political economy.

Production

The decomposition of formerly national systems of production and their re-articulation across international boundaries has marked the current phase of globalization. Unfettered and mobile capital seeks out cheap and flexible labor around the globe. This has been most marked in manufacturing, but it is increasingly true of service provision too. Not only are customer services and back-office work sent offshore, but Northern capital seeks increased returns from providing services to locals in the target country, whether the broad population of the developed nations, or the new middle classes of the developing world (UNCTAD 2002). For example, provision of health care, from hospitals to diagnostic teams, comes increasingly from external sources. This investment is facilitated by removing barriers to the free movement of capital into and out of states and by privatizing public assets. The *General Agreement on Trade in Services* (GATS) of the World Trade Organization (WTO) commits states to allowing unrestricted inward investment and full repatriation of profits by non-national service providers.[2] The transnationalization of health care provision benefits still more directly from the work of the World Bank which actively invests in private medical businesses in countries such as India and Brazil.[3]

Consumption

Capital invested requires a return. That is only possible where there is a functioning market with effective demand for private medical services. Under the prevailing neo-liberal order, where state funders will not cover privately provided treatment, patients should be left with enough income after taxation to pay for it themselves. This is increasingly the case in most nation states: fiscal constraints, driven by fear of investment strikes, have the dual effect of degrading public services and freeing up private resources (Leys 2001, 81). The erosion of public services is furthered by rhetoric disparaging state provision as irredeemably inefficient and inadequate. Private providers and senior professionals strive to make a plausible case for privatization in the name of choice and quality.

Consumption of health care, just like its provision, is no longer confined by national borders. Again GATS is set to aid this process. It requires states not only to allow foreign service providers in, but also to permit their own nationals to travel to avail of services abroad and to export sufficient funds to pay for this.[4] As will be seen these developments have been anticipated in the law of the European Union.

Where the service cannot come to the consumer, the consumer is to be assisted in journeying to the service. Capital flight, once seen as the bane of third world development, is now enshrined as a right in international and domestic law, enjoyed by the health care industry and its wealthy clients (Adelman and Espiritu 1993).

Thus, networks both of production and consumption are established. A global market is being constituted as transnational service providers attract nomadic patient-consumers. Economic globalization, driven by the relentless quest for profit of corporations in the developed countries and enforced by international economic law, acts inevitably to decompose the bounded and solidaristic basis of national health care systems. In the UK, for example, the NHS monopoly of provision has been broken up. Foreign as well as domestic companies now contract with the UK government for the provision of services (Pollock 2004). In developing countries an expanded market for private health insurance and cherry-picking by the relevant companies draws the upper and middle classes away from the state system. The national risk pool of patients disintegrates and the poor and lower middle classes are thrown upon an under-funded rump system of public health care.[5]

Commodification and the "New Medicine"

Another dynamic feature of the contemporary scene is the development of what has been called the "new medicine": organ transplantation, assisted reproduction and human genetics (Richardson and Turner 2002). Body parts (e.g. kidneys) and particles (e.g. stem cells) are the basic material of these therapies. Demand for them has opened up new opportunities for "primitive accumulation" or "accumulation by dispossession." The latter term originally described the often violent phase of expropriation preceding more orderly regimes of capitalist accumulation. Thus, during the industrial revolution in Britain peasants were dislodged from their smallholdings by "reforming" landowners and compelled by economic necessity to seek waged work in the new factories.

However, this was not a one-off event. Primitive accumulation has remained a feature of capitalism up to the present globalized era (Harvey 2003). As Rosa Luxemburg (2003, 397) put it,

> Historically, the accumulation of capital is a kind of metabolism between capitalist economy and those pre-capitalist methods of production without which it cannot go on and which in this light it corrodes and assimilates.

Seeking an outlet for investment, a market for its products and a source of labor and raw material, capitalism has always been forced beyond its own geographic and social limits. Imperial conquest has been interpreted in this way (Arendt 1968). In the current era capitalism continues to cross the frontiers separating it from non-market realms, such as the welfare state and its third world counterpart the developmental state. It also penetrates the taboos surrounding the human body: commodifying organs, human tissue and genetic material (Leibovitz-Dori 1997). These are acquired for money as inputs in the production of health care. Their "processing" (e.g. through transplantation) creates further value which is realized in the form of fees earned for the service.[6]

The introduction of means of transport, such as railways, was vital to the spread of the commodity economy under pre-First World War imperialism.[7] Similarly the extension of advanced Western medical technology is essential to the accumulation of capital in the health care sector. Standard techniques and internationally valid protocols make for a uniform medicine practicable across the globe (Mol and Law 1994). They enable the primary produce of the new medicine to be extracted and to circulate in the global market. As Britain's nineteenth century Opium Wars show, the introduction of the commodity economy has often been far from peaceful. Similarly the marketization of human organs and the depletion of public health care provision has not gone uncontested. In particular, resistance to structural adjustment and privatization programs has been sporadic, but often intense.[8]

Globalized localism – Localized globalism

Boaventura de Sousa Santos offers a further, useful perspective on globalization which may be adapted to health care (de Sousa Santos 2002, 182). He argues that there is no such thing as a pure globalism. What we encounter are in fact "globalized localisms": the practices of a specific state or region which extend over the globe, gaining the power to define their rivals as "merely local." The asymmetric relationship between scientific Western medicine and the traditional therapies of African peoples is a good example.[9] Globalized localisms find their counterparts in "localized globalisms." Just as the former cannot be understood as abstractly universal, so the latter do not correspond to the merely particular. Localized globalism connotes instead the specific impact of transnational practices and imperatives on local conditions. The enforcement in African jurisdictions, at a time of public health crisis, of patents held by American and European pharmaceutical companies provides an instance of this (Nagan 2002).

The pattern which Santos describes is significantly conditioned by the historic inequalities of the world system.[10] The different trajectories to modernity of different countries determine their relative positions in this system. The former metropolitan powers of Europe, as well as the settler nations of North America and Australasia, form the core; the former colonies of Africa, South Asia and Latin America, the periphery. It is argued that East Asian nations, with their commonly autarkic route to modernity, are moving from the former to the latter. Santos points out that countries at the core specialize in producing globalized localisms; those at the periphery are commonly forced to bear the costs of localized globalisms (de Sousa Santos 2002, 179). Since ours is a capitalist globalization, this polarity can be represented as a hierarchical division of labor on a world-scale. The specific practices of capitalist industrialism, service provision, financial governance and legal ordering[11] are exported from the strong states as globalisms, to be localized in the weaker states reshaping their material and normative prospects.

Latterly the achievements of the core nations have been mediated through institutions of global governance such as the World Bank, the International Monetary Fund (IMF) and the WTO. These compel developing countries to "reform" (i.e. privatize) their public sectors and to implement the "rule of law" within their territories.[12] Yet such programs have their origins in the practices and reforms of specific Western nations. Indeed their implementation in the developing world boosts invisible export earnings by developed country academics, NGOs, civil servants and management consultants (Sassen 2003; Wallace 2003). It is, of course, incorrect to view any developing country as the undifferentiated recipient of external diktats. A substantial section of the ruling group will be active as the local steward of globalization.[13] A cadre of bureaucrats will identify with the reform project and ensure its legislative and administrative implementation. Hosts of NGOs take over the state's welfare functions and answer directly to foreign agencies for the expenditure of grant monies (Albo 2003). Localized globalisms take effect, therefore, not simply in material terms. They also reshape social and political structures within developing countries.

The following may serve as an example.[14] British health economists, working within the paradigm of that discipline in the late 1980s, develop models of health care funding. In particular they recommend the imposition of "user fees" on patients to discourage unnecessary use of facilities (Lawson 1994). This is the "localism." It achieves the status of "globalism" through the powerful agency of the World Bank. The Bank adopts "user fees" as part of its strategy for promoting efficiency in public health services.[15] It imposes the policy on developing countries, such as Tanzania, as a condition of further loans (Kiwara 1994). The policy is adopted into Tanzanian law, and implemented by officials at the ministries of finance and health. They are advised by British academics and civil servants. The policy is experienced as a localized globalism by existing users of clinics around the country. They bear its costs, refraining from necessary use of health facilities and suffering an increase in conditions such as anemia, seemingly as a result (Hussein 1995; Hall 2001).

Uneven normative convergence

Chase-Dunn has argued that the capitalist world economy is integrated more by political-military power and market interdependence than by normative consensus (Chase-Dunn 1991, 88). Of course arguments are made for both new and revived normative universalisms either functionally, in response to economic globalization, or on a priori grounds.[16] But their realization, thus far, has been only partial in geographic and sectoral terms (de Sousa Santos 2002, 171). This is of especial significance in the area of medical law. Legal commentators routinely use the canon of Western ethics as a meta-discourse for the articulation and resolution of regulatory problems.[17] Yet the historically contingent and geographically specific pedigree of Kantianism, utilitarianism and so on is obvious; as is the lack of consensus on a range of substantive issues, like abortion or the right to health care. In fact it is argued that, far from being mere "survivals," normative and cultural particularisms are adaptive responses to economic globalization (Amin 1998). As states withdraw from productive and welfare activities which ameliorate the effects of the free market, legitimacy is renewed via ethnic, nationalistic and religious mobilizations (Betz 2002). These can on occasion accentuate differences in the legal treatment of ethically sensitive medical issues. However, the dialectical progress of capitalist globalization means that these legitimation strategies are undermined at the same time as they are promoted by commodification and the decline of pre-capitalist social structures. We shall see that precisely this has been true of the commodification of organ sales in the developing world.

Norms are not absent from capitalist globalization. No matter that the chain of production and consumption now crosses multiple borders, value is still created and realized within the territorially defined jurisdictions of nation states. Orderly accumulation, thus, requires the stability provided by a dependable and suitably oriented national system of contract, property, labor and commercial law. The global moment of this legal regime is found in the normative output of the WTO, the World Bank and the IMF, as well as regional bodies such as the European Union (EU). Treaty obligations (e.g. GATS) and loan conditionalities, backed up by formal and informal sanctions, compel nation states to develop and maintain essentially similar pro-market legal regimes.[18] Broad convergence on privatized health care and the global protection of pharmaceutical company rents are the fruit of national legislation mandated by international economic law (Shaffer and Brenner 2004). We find normative consensus to be most advanced where it most intimately regulates and protects the globalized system of accumulation. Legal harmonization or unification proves more difficult to achieve in non-economic sectors or where the issue cannot be reformulated as a matter of economic liberties.[19]

ENGLISH MEDICAL LAW AND THE MARKET

In the period from 1945 until the early 1970s, economic struggles within nation states concerned how the social product would be distributed, as between labor and capital (Harvey 2003, 77). Their growing intensity in the 1970s reflected the declining profitability of companies in the Western countries (Offe 1982). These struggles were commonly centered on the workplace, but they also found a limited outlet in litigation to compel governments to maintain and expand welfare provision. The general crisis of the 1970s was resolved through liberalizing transnational capital flows in the manner discussed above. With labor decisively weakened by job insecurity and state compulsion, contemporary social struggles are now more likely to involve the defense of natural endowments, traditional knowledge and extant systems of public welfare, as well as the valorization of minority identities and lifestyles. With the rise of human rights law, the effects of this capitalist globalization are increasingly felt in the courts. Disputes about intellectual property in lifesaving drugs, attempts to hold private health care providers to account and struggles over the commodification of traditional knowledge have marked out the new medical law jurisprudence.[20]

These changes can also be tracked in the case law of the English courts. Since the foundation of the National Health Service in 1948, English health care law has been shaped by a number of key assumptions regarding the nature and aims of medical work in a state-funded and publicly delivered system.[21] These effectively created a zone of professional autonomy within which the medical profession was allowed to deliver health care free from the compulsions of the market and the demands of patients. Judges routinely deferred to clinical judgment in decisions on medical malpractice and in adjudicating the health care entitlements of patients denied access to treatment (Brazier 2003). The standard of information disclosure was determined by medical expert opinion, not by the patient's right to self-determination (Jones 1999, 103-34). Legislation permitting abortion was passed in 1967, but access to termination would ultimately depend on clinical judgment and not women's rights.[22] There was a further presumption, embodied in a range of legislation, that the human body should not be commodified. Organ trafficking and commercial surrogacy were prohibited.[23] In judicial rhetoric, as well as statute law, doctors were thus valorized as the near-sovereign custodians of a precious and scarce national resource. This image was informed by an inherited Victorian prejudice in favor of the doctor as gentleman practitioner and a faith in the medical profession as the agent of social progress.[24]

The effect of this ideological formation in law was to insulate doctors from external scrutiny. It also served to conceal behind a veil of clinical discretion the increasingly acute rationing implemented under neo-liberalism from the mid-1970s on. Challenges to the postwar orientation of medical law have taken two main forms. The first, resting on a "social critique,"[25] have sought the redistribution of health care or general resources toward favored medical causes. While legal challenges are necessarily individual, they summate to a demand for increased funding of the National Health Service. As has been noted, such challenges are generally rejected on grounds of justiciability (Whitty 1999). The scarcity of health care resources is naturalized, a matter of fate which no judge could set right. The second set of challenges, resting on an "artistic critique," has sought the emancipation of patients from the patriarchal dominance of medical practitioners.[26] The infusion of human rights discourse and bioethics into medico-legal practice testifies to the success of this critique. These challenges have also re-valorized market models of health care, even where they were originally inspired by the anti-capitalist movements of the late-1960s.[27] Thus, the radical demand for patient autonomy can be seen as a justification for increasing patient choice and the adoption of market systems in the delivery of health care (Jacob 1988). In the following two sections we shall examine the effects of these tendencies: a re-commodification of medical practice, with the patient role transformed from passive recipient to active, mobile consumer.

INTERNATIONAL ORGAN TRAFFICKING AND ENGLISH LAW

Under the UK *Human Tissue Act* 2004 a punishment of up to three years imprisonment may be imposed on persons giving or receiving rewards for the supply of organs, seeking to find others willing to supply organs or managing a company involved in the negotiation or initiation of organ sales.[28] Publication and distribution of advertisements in this connection are punishable by up to fifty-one weeks imprisonment.[29] To this extent the new Act continues the explicit ban on organ trading first introduced into British law by *Human Organ Transplantation Act* 1989.[30] The latter was passed in response to a scandal involving the extraction of organs from Turkish men for the benefit of British patients. The 2004 Act, thus, reinforces the general orientation of English medical law towards non-market values: in this case the taboo against commodification of the human body. Restrictions on payment for surrogacy arrangements, blood donation, and the supply of human gametes are consistent with this. The ethic of altruism founded on "gift relationships" remains at the ideological heart of health care law in Britain.[31] Similar measures have been enacted by most other developed[32] and indeed many less-developed nations.[33] At a global level UNESCO (UNESCO 1989), the World Health Organization (WHO 1991) and the World Medical Association (WMA 2000) are all opposed to the creation of markets in organs.

Notwithstanding these measures organ trafficking continues to grow. An exact quantification is, of course, impossible. Nonetheless the anecdotal evidence is strong (for examples, see: Scheper-Hughes 2003). The extensive development of illegal organ markets in, for example, India is well documented (Goyal et al. 2002). In the UK a number of doctors have been disciplined by the General Medical Council for performing "broker" functions, creating markets for the Indian transplantation business (BBC 2002). A number of reasons for this growth, linked to the preceding discussion of globalization, can be suggested.

1.   Globalization of health care production, as well as the falling costs of transport. Western patients can travel relatively cheaply to countries such as Turkey or the Philippines. They stay in hospitals of a Western standard and receive treatment at least as good that in their home countries.[34]

2.   Effective demand on the part of wealthy patients is met by supply from locals sufficiently desperate to undergo the risks of operation and the removal of organs. Market impediments are easily circumvented. In fact, legal prohibitions themselves become the objects of a parallel market in bribes and favors.

3.   Since the 1980s immunosupressant drugs have greatly increased the success rate for transplantation. Usable body parts are now available for circulation in the international medical market.

These developments have begun to erode the taboo against commodification at national level, in Britain and elsewhere. While the *Human Tissue Act* 2004 has maintained the existing prohibition, there are signs elsewhere of changing attitudes.

The British Medical Association has, for example, recently hosted a widely reported debate on the question, an event unthinkable just ten years ago.[34] Bioethicists and other moral philosophers address the justifiability of organ trading in growing numbers (Radcliffe Richards et al. 1998; Wilkinson 2003). The great majority support some kind of regulated market. In their arguments technical feasibility, unmet demand and untapped supply coalesce into a moral defense of organ sales. Scarcity is taken to be a natural phenomenon rather than a product of conscious choices to invest in transplantation facilities and, on a global scale at least, to privilege the lives of a wealthy minority (Lock 2002, 1406). Proponents of markets dismiss taboos against commodification as indefensible in liberal and pluralistic societies (Duxbury 1996). They concentrate instead on the possibility of impaired consent on the part of the organ seller. They find it "hard to see how an offer of money *per se* could constitute illegitimate pressure" in an organ transaction (Herring 2002, 53). This is of course unarguable; even in cases of economic necessity it can be shown that the consent of the seller was "real."

However, there are two significant and related lacunae in pro-market arguments. First, global issues are usually bracketed in these discussions. Proposed markets are limited to a single state or a developed region such as the European Union (Erin and Harris 1994). Conditions in developing countries are too extreme to permit any direct extension of the pro-market argument. In spite of the growing significance of transnational organ tourism, ethicists are thus often self-restricted to the national horizon. Second, writers in this vein foreground agency over structural concerns. The latter are addressed, if at all in fatalistic terms. For example, Pattinson (2003) makes the valuable, but regrettably underdeveloped point that

> [Exploitation and inequality of bargaining power] are not restricted to commercial organ dealings. Many labor markets, especially in the developing world, pay workers paltry sums of money...It is difficult to see why if these concerns justify prohibiting organ dealings, rather than the need for regulation and supportive structures, they do not justify the prohibition of any activity paying low wages and generating large sums of money.

Regulation can indeed improve the likelihood and the quality of consent obtained from organ sellers. But it is itself dependent on the political and economic context in which it must operate. This context is, as has been discussed, decisively shaped by international relations that reproduce economic and political inequalities between different states, and within states (Scheper-Hughes 2001, 62). The ethics of organ markets inevitably implicate questions of social and global justice not readily fitted within the analytical grid of liberal bioethics.

What are the structural issues raised by organ tourism between developed and developing countries? On examination we find that many of the problems associated with other forms of commodity production and trade can be expected here too.

1. Continued direction of resources toward intensive production (here hospital medicine), which benefits consumers in the North, and away from interventions aimed at the majority of people in the South (here basic public health).

2. Increased threat to the livelihood and indeed the lives of poor people posed by their participation in commodity production. A peasant who favors cash crops over subsistence is more exposed to a falling market. Where a poor person sells a kidney, their capacity to labor and earn is permanently vulnerable to further illness.[35]

3. Replication and exaggeration of divisions internal to the particular state or region. Class, gender and ethnic inequalities are commonly reinforced when articulated with the imperatives of the global economy.[36] At the margins of the global economy a "transplantation underclass" is already in place. Depending on the particular region, it is composed of poor women, displaced peasants, the homeless, prisoners and the mentally ill.[37]

4. At present supply often meets demand for organs as result of economic coercion, fraud or physical force.[38] Yet the institutions which might provide for fairness in the market are often dysfunctional, bankrupt or corrupt. The hollowing out of the state under structural adjustment programs, and the correlative rise of competitive markets in formerly public services has diminished local regulatory capacity.

While systems for extracting and marketing organs have been successfully, if often illicitly reconstituted at global level, there has been no matching ethical and cultural convergence. The strength of taboos relating to organ removal still varies considerably as between countries and regions (e.g. strong in Japan, less so in India) (Lock 2002, 1412). Enforcement capacities differ too. Furthermore the national consensus against commodification has come apart under pressure of the actually-existing market. Legal bans on trafficking, such as that in Britain, are attacked in principle and contradicted in practice. Despairing of their ability to protect the vulnerable through prohibitions, commentators prefer to settle for a lesser evil within the unchallenged horizons of global inequality and structural exploitation (Friedlander 2002). Arguing at what Santos has called the sub-paradigmatic level, they urge reform and adaptation rather than contest and transformation (de Sousa Santos 2002, 173).

HEALTH TOURISM IN EUROPE

Health tourism for more routine procedures is also increasing within the developed world. We have already noted that this poses threats to the largely solidaristic basis of national health care systems, whether insurance-based, as in continental Europe, or state-funded and run like Britain's NHS. Mobile patients draw off resources from the national system restricting the ability of local providers to maintain and expand capacity. Though the cost of air travel and medical procedures across the globe is falling, private health tourism remains out of reach for most citizens, even in the developed world. Increasing effective demand will only be possible if state health insurers and providers are willing to fund cross-border treatment. The WTO's *General Agreement on Trade in Services* has already been mentioned as an impulse to the creation of a global health care market in this way. At the regional level, European Union law has in recent years been interpreted to facilitate the mobility of patients between member states (Hervey and McHale 2004). The impact of intra-European free trade rules on Britain's NHS has recently been made clear in the case of *Secretary of State for Health v. R (on the Application of Watts).*[39]

Mrs.Yvonne Watts had been told to wait twelve months for a total hip replacement operation by her local Primary Care Trust (PCT). Since this was within the Department of Health's target waiting time of fifteen months the PCT refused to fund a trip to Lille in France to have the operation performed there at an earlier date. She proceeded at her own expense. Seeking judicial review of the PCT's refusal,[40] Mrs. Watts invoked her European Community law right to travel to avail of services provided in another member state. In implementation of this right, she claimed, the PCT was obliged to reimburse her costs. At first instance, Munby J held in her favor on the point of Community law. On the facts, however, it appeared that the PCT had made a revised offer of treatment two months before the scheduled date of the Lille operation which Mrs. Watts could reasonably have accepted. Her claim failed accordingly.

The Secretary of State for Health appealed against the ruling that, on principle, there was a right to reimbursement. Lord Justice May for the Court of Appeal ultimately held that decision of the case should be suspended and a reference made to the European Court of Justice (ECJ) for clarification of the law. The Luxembourg court had not yet ruled in this case at the time of going to press. It is nonetheless worth studying the reasoning of the Court of Appeal and its reflections on the health policy implications of the case. The situation of the English Court is seen to be a poignant one, on the brink of a decisive rearrangement of the value hierarchy in this area of medical law. The practical implications of this re-ordering for the form and extent of public health provision in the United Kingdom are likely to be profound.

Article 49 (ex 59) of the European Community Treaty prohibits restrictions on the freedom to provide services to nationals of other member states. The ECJ has erected a substantial edifice of interpretation on Article 49 (ex 59), to the extent that appeals to its literal meaning "may not be regarded as persuasive."[41] Thus, the right of a consumer to travel to avail of services has been guaranteed as a corollary of Article 49 (ex 59).[42] This effective right is further expanded by Article 22 of Council Regulation 1408/71, which provides that a recipient of social services in one member state may avail of "treatment appropriate to his condition" in another member state at the expense of the relevant home institution. The ECJ has defined "home institution" to include state-backed contributory sickness funds.[43] Funding may be refused by the home institution unless the patient cannot be offered treatment "within the time normally necessary for obtaining" it at home.[44] The issue in *Watts* was whether the refusal of the relevant authorization by the PCT and the Department of Health was supported by the exception to Article 22. In other words was it an objectively justifiable and proportionate restriction on Mrs. Watts' Article 49 (ex 59) rights?

In the terms of the relevant ECJ jurisprudence, the Court of Appeal had to decide whether there would be "undue delay" in treating Mrs. Watts if she were not enabled to undergo the operation in France.[45] The Secretary of State contended that the treatment-specific waiting list times prescribed for PCTs by the Department of Health should be taken into account in this decision. It argued that objective justification for a restriction of this scope was provided by the need for financial stability in public health care systems. While recognizing that this indeed was the broad justification for the authorization requirement contained in Article 22, the Court of Appeal held that national waiting lists could play no role in determining the question of "undue delay." The ECJ had established, most recently in the case of *Inizan*,[46] that the "time normally necessary" for obtaining treatment is solely a matter of clinical judgment. The extent of the patient's disability, their pain and likely prognosis were the co-ordinates of this judgment, exclusive of the detailed economic considerations embodied in waiting lists.

May LJ thus followed the logic of the ECJ to its conclusion, namely that whenever a patient's doctors judged them to be in need of treatment sooner than the waiting time prescribed by the Department of Health, that patient should be entitled to "jump the queue" by traveling to another member state with the financial support of their local PCT. He was plainly disturbed by the prospects for the National Health Service opened up by this entitlement. In particular he was sympathetic to the argument that the effect of Community law here would be to

> disrupt NHS budgets and planning and undermine any system of orderly waiting lists...
> [Furthermore] if the NHS were required to pay the costs of some of its patients having
> treatment abroad at a time earlier than they would receive it in the United Kingdom this
> would require additional resources.[47]

Since waiting lists were a product of scarce resources this extra funding could only be obtained if

> those who did not have treatment abroad received their treatment at a later time than
> they otherwise would or if the NHS ceased to provide some treatments that it currently
> does provide.[48]

His decision to refer to the ECJ a set of issues which had been largely settled in earlier cases testifies to his concern for clarity in an area of constitutional significance for the NHS. If as the ECJ has held, "Community law does not detract from the power of member states to organize their own social security systems,"[49] can it be true that the edifice constructed on Article 49 (ex 59) EC operates "to dictate the national health service budget of the individual member states?"[50]

CONCLUSION

This essay has examined some of the implications of globalization for the content of English medical law. As a field of academic and popular discourse, as well as practical decision making, the latter was constituted by a set of anti-market exclusions and prohibitions. An ethos of altruism pervaded the self-understanding of the medical profession, and its representation in law. With health care free at the point of use the image of the doctor as a selfless servant of the greater good was realized in daily practice. Market transactions at the margins of standard medical care were also prohibited or strictly limited. Altruistic medicine was at the same time patriarchal medicine. Legal exclusion of the market from British health care was reinforced by a notable limitation of patients' rights. This paper rested on the central assumption that this dispensation in medicine and law was intimately connected to the distinctive political and economic conjuncture of the postwar decades.

If globalization is defined as a set of profound changes in political economy, then we must accept that the conjuncture which supported patriarchal altruistic medicine in Britain has been dissolved. We have attempted to sketch some features of the new conjuncture: increasingly marketized health care; rights consciousness among patients; cross-border consumption of health care; and the weakening of solidaristic national health systems. The leading subjects of the old dispensation were the state, acting on behalf of the masses, and the medical profession. The leading subjects of the new dispensation are commercial health care providers and their paying customers. If postwar medicine in Britain had some of the features of a feudal society organized by rank, then the effect of globalization has been to bring the bourgeois revolution to this discrete sector of social relations.

NOTES

I am grateful to Catriona Sangster for research assistance, and to Ambreena Manji for comments on an earlier draft. Responsibility for errors is mine alone.

1. I use English law throughout to refer the common law of England and Wales, as distinct from the civilian law of Scotland.
2. GATS Article XI. The agreement inaugurates successive rounds of negotiations aimed at liberalizing different sectors of service provision, rather like the General Agreement on Trade and Tariffs (GATT) did for trade in goods. Banking and financial services have already been subject to negotiations, but not yet health care. Commentators have argued, however, that GATS will have a profound indirect effect on health care provision in any case given the many different modes of delivering that service (Price, Pollock and Shaoul 1999).
3. In 1999, for example, the World Bank invested $8 million in a 270-bed private hospital in Calcutta, owned and run by Singapore-based Parkway Enterprises (Hall 2001, 9).
4. GATS Articles I.2b and XI.1 respectively.
5. This process has been noted of Chile in the period since the coup of 1973. Under General Pinochet's free market "reforms" the comprehensive national health care system was broken up into a residual public service and a number of competing private insurers and providers. Effectively the wealthiest 30 percent of the population were freed of social responsibilities with regards to heath and allowed to purchase the kind of "high-tech" care discussed here (Collins and Lear 1995).
6. Thus, the University of Pittsburgh proposed in the mid 1990s to exchange its transplantation expertise for a supply of surplus livers from hospitals in Sao Paolo, Brazil (Scheper-Hughes 2000). Harvard Medical School has joined with the World Bank and an Indian pharmaceutical company to train heart specialists at a private hospital in the state of Maharashtra (Sexton 2001).
7. The construction of the railway from Mombasa to Kampala in the 1890s is emblematic of this, see Hill (1976).

8. As well as the convulsion of many Latin American countries, the last number of years have seen a localized refusal of the ANC's neo-liberal programme in formerly loyal districts of South Africa, see Ngwane (2003).

9. Thus, s 41 of the *Tanganyika Medical Pracitioners and Dentists Ordinance*, passed by the British colonial government in 1948 and still in force today, states that "native medicine" may only be practiced upon members of a traditional healer's own ethnic group. By contrast Western medicine, regulation of which is the chief object of the Ordinance, is implicitly unrestricted in this respect. The definition of the particular and its subordination to a universal is achieved through law, and in the context of a colonial project itself one of subordination and peripheralization. The interactions of traditional healers, local regulators and multinational bioprospectors in modern Tanzania are thus decisively shaped by a distinctive history of globalization.

10. At this stage of his account Santos relies on Wallerstein (1984).

11. Respective examples might be "Taylorized" factory discipline, the customer service "ethos" of telephone banking facilities, constitutionally-anchored independence of central banks and individual titling of rural landholdings.

12. For the perspective of a critical insider, see Stiglitz (2002).

13. On the question of class formation under globalization in developing countries, see Burnham (2002).

14. It is drawn from Harrington (1998).

15. For a recent statement, see World Bank Development Report (2003, 71).

16. As regards the field of health law and policy, an interesting case for "communitarian globalism" is made by Aginam (2000).

17. For a sensitive posing of the issues, see Singer (2004).

18. For a detailed view of the international institutions concerned with health law, see Koivusalo and Ollila (1997).

19. For a compelling exploration of this effect, see Fidler (2004).

20. For an overview, see Fort, Mercer and Gish (2004).

21. I have examined these elsewhere (Harrington 2002; 2004).

22. The agreement of two registered medical practitioners is required before any termination is lawful: *Abortion Act* s 1(1) (Sheldon 1997).

23. See respectively, *Human Organ Transplantation Act* (1989) s. 1; *Surrogacy Arrangements Act* (1985) s.2.

24. For discussion of the origins of these images, see Lawrence (1994).

25. I take this idea, and the companion notion of an "artistic critique" from Boltanski and Chiapello (1999).

26. For a path breaking formulation of this critique, see Kennedy (1980).

27. The original leftist fusion of "social" and "artistic" critiques is demonstrated in the medical context by Doyal (1979).

28. *Human Tissue Act* 2004 s 32(1).

29. *Human Tissue Act* 2004 s 32(2).

30. The 1989 Act itself has been repealed: see *Human Tissue Act* 2004 ch 7.

31. For an important ethical defence of this, see Titmuss (1970).

32. For a recent summary of European legislation, see Steering Committee on Bioethics and European Health Committee (2003).

33. In India, for example, a leading destination for "organ tourists": *Transplantation of Human Organs Act* 1994 s 19.

34. See papers delivered at the conference "Medical Ethics Tomorrow" held in London, 3rd December 2003. Available at: <http://www.bma.org.uk/ap.nsf/Content/MedicalEthicsTomorrowConfPapers> (Last accessed: 31 March 2005).

35. Since the supply of healthy organs vastly exceeds demand, the returns to sellers will be very low in the long run, with most profits being extracted by brokers (Kolnsberg 2003).

36. Reports from India allege that the market in organs has breathed new life into the practice of dowry. "Women flow in one direction, kidneys in the other" (Scheper-Hughes 2000, 26).

37. In many countries contemporary organ markets build on earlier more coercive extraction programmes under authoritarian regimes: in Argentina the right-wing military targeted political

progressives; in South Africa most "donors" were young blacks, most recipients older whites (Scheper-Hughes 2000, 26).
38. To take one example, the premier of the Indian Punjab accused a group of doctors, lawyers, brokers and politicians of operating a $35 million organ trade in the state causing the deaths of 24 "donors." Police reported that most victims were poor migrant labourers from Northern and North Eastern India (West Australian 2003).
39. [2004] EWCA Civ 166.
40. *R v. Bedford Primary Care Trust, ex parte Watts* [2003] EWHC 2228.
41. *Secretary of State for Health v. R (on the Application of Watts)* [2004] EWCA Civ 166 (Per May LJ at para 31). See further: van der Mei (2003).
42. *Luisi and Carbone v. Ministero del Tesoro* [1984] ECR 377.
43. *VG Müller-Fauré v. Onderlinge Waarborgmaatschappij OZ Zorgverzekering UA* (C-385/99), 13 May 2003 unreported; *Geraets-Smits v. Stichtung Ziekenfonds VGZ* [2001] ECR I-5473; *Vanbraekel v. Alliance Nationale des Mutualités Chrétiennes* [2001] ECR I-5363.
44. Regulation 1408/71 article 22(1)(c).
45. *Inizan v. Caisse Primaire d'Assurance Maladie des Hauts-de Seine* (C-56/01).
46. Ibid.
47. *Secretary of State for Health v. R (on the Application of Watts)* [2004] EWCA Civ 166 (Per May LJ at para 105).
48. Ibid.
49. *Geraets-Smits v. Stichtung Ziekenfonds VGZ* [2001] ECR I-5473 (at para 44). In this the ECJ is following Article 152.5 EC.
50. *Secretary of State for Health v. R (on the Application of Watts)* [2004] EWCA Civ 166 (Per May LJ at para 107).

REFERENCES

Adelman, S., and C. Espiritu. 1993. The debt crisis underdevelopment and the limits of law. In: *Law and crisis in the Third World*, eds. S. Adelman, and A. Paliwala, 172-94. London: Hans Zell.
Aginam, O. 2000. Global village, divided world: South-North Gap and global health challenges at century's dawn. *Indiana Journal of Global Legal Studies* 7: 603-45.
Albo, G. 2003. The old and new economics of imperialism. In: *The new imperial challenge. Socialist Register 2004,* eds. L. Panitch, and C. Leys, 88-113. London: Merlin Press.
Amin, S. 1998. *Capitalism in the age of globalization. The management of contemporary society.* London, New York: Zed Books.
Arendt, H. 1968. *Imperialism.* New York: Harcourt Brace.
BBC. 2002. Organ trade GP struck off. 30 August. Available at: <http://news.bbc.co.uk/1/hi/health/2225357.stm> (Last accessed: 11 April 2005).
———. 2002. Organ trade GP suspended. 15 October. Available at: <http://news.bbc.co.uk/1/hi/health/2329447.stm> (Last accessed: 11 April 2005).
Betz, H.G. 2002. Xenophobia, identity politics and exclusionary populism in Western Europe. In: *Fighting identities: Race religion and ethno-nationalism. Socialist Register 2003,* eds. L. Panitch and C. Leys, 193-210. London: Merlin.
Boltanski, L. and E. Chiapello. 1999. *Le nouvel esprit du capitalisme.* Paris: Gallimard.
Burnham, P. 2002. Class struggle, state and global circuits of capital. In: *Historical materialism and globalization,* eds. M. Rupert, and H. Smith, 113-28. London, New York: Routledge.
Brazier, M. 2003. *Medicine, patients and the law.* Harmondsworth: Penguin.
Chase-Dunn, C. 1991. *Global formation: Structures of the world economy.* Cambridge: Polity Press.
Collins, J., and J. Lear. 1995. *Chile's free market miracle: A second look .* Oakland: Food First Books.
de Sousa Santos, B. 2002. *Toward a new legal common sense: Law, globalization and emancipation.* London: Butterworths.
Doyal, L. 1979. *The political economy of health.* London: Pluto.
Duxbury, N. 1996. Do markets degrade? *Modern Law Review* 59: 331-48.

Erin, C.A., and J. Harris. 1994. A monopsonic market: Or how to buy and sell human organs, tissues and cells ethically. *Fulbright Papers* 15: 134-53.

Fidler, D. 2004. Germs, norms and power: Global health's political revolution. *Law Social Justice and Global Development* 7. Available at: <http://elj.warwick.ac.uk/global/issue/2004-1/fidler.html> (Last accessed: 11 April 2005).

Fort, M., M.A. Mercer, and O. Gish, eds. 2004. *Sickness and wealth. The corporate assault on global health*. Chicago: South End Press.

Friedlaender, M. 2002. The right to buy or sell a kidney: Are we failing our patients? *The Lancet* 359: 971-3.

Giddens, A. 1998. *The third way: The renewal of social democracy*. Cambridge: Polity.

Goyal, M., R.L. Mehta, L.J. Schneiderman, and A.R. Sehgal. 2002. Economic and health consequences of selling a kidney in India. *Journal of the American Medical Association* 288: 1589-93.

Hall, D. 2001. *Globalisation, privatisation and healthcare*. London: Public Services International Research Unit.

———. 2001. *Globalisation, privatisation and healthcare – A preliminary report*. Greenwich: Public Services International Research Unit. Available at: <http://www.psiru.org/reports/2001-02-H-Over.doc> (Last accessed: 18 April 2005).

Harrington, J. 1998. Privatizing scarcity: Civil liability and health care in Tanzania. *Journal of African Law* 42: 147-71.

———. 2002. "Red in tooth and claw": The idea of progress in medicine and the common law. *Social and Legal Studies* 11: 211-32.

———. 2004. Elective affinities. The art of medicine and the common law. *Northern Ireland Legal Quarterly* 51: 259-76.

Harvey, D. 2003. The "new" imperialism: Accumulation by dispossession. In: *The New Imperial Challenge: Socialist Register 2004*, eds. L. Panitch, and C. Leys, 63-87. London: Merlin.

Henwood, D. 2003. *After the new economy*. New York, London: New Press.

Herring, J. 2002. Giving, selling and sharing bodies. In: *Body lore and laws,* eds. A. Bainham, S. Day-Sclater, and M. Richards, 43-62. Oxford: Hart.

Hervey, T.K., and J.V. McHale. 2004. *Health law and the European Union*. Cambridge: Cambridge University Press.

Hill, M.F. 1976. *Permanent way. The story of the Kenya and Uganda railway*. Nairobi: East Africa Literature Bureau.

Hussein, A.K. 1995. *The Effect of User Charge Policy and Other Non-Price Factors on the Utilization of Health Services in the Dar es Salaam region –1994 (*Dissertation submitted in partial fulfilment of M.Med (Community Health) degree University of Dar es Salaam 1995).

Jacob, J. 1988. *Doctors and rules. A sociology of professional values*. London: Routledge.

Jones, M.A. 1999. Informed consent and other fairy stories. *Medical Law Review* 7: 103-34.

Kennedy, I. 1980. *The unmasking of medicine*. London: Allen & Unwin.

Kiwara, A.D. 1994. Health and health care in a structurally adjusting Tanzania. In: *Development strategies for Tanzania. An agenda for the twenty first century*, eds. L.A. Msambichaka, H.P.B. Moshi, and F.P. Mtatifikolo, 269-90. Dar es Salaam: Dar es Salaam University Press.

Koivusalo, M., and E. Ollila. 1997. *Making a healthy world: Agencies, actors and policies in international health*. London, New York: Zed Books.

Kolnsberg, H.R. 2003. An economic study: Should we sell human organs? *International Journal of Social Economics* 30: 1049-69.

Lawrence, C. 1994. *Medicine in the making of modern Britain, 1700-1920*. London, New York: Routledge.

Lawson, A. 1994. *Underfunding in the social sectors in Tanzania: Origins and possible responses*. Dar es Salaam: TAREG.

Leibowitz-Dori, I. 1997. Womb for rent: The future of international trade in surrogacy. *Minnesota Journal of Global Trade* 6: 329-54.

Leys, C. 2001. *Market-driven politics. Neoliberal democracy and the public interest*. London: Verso.

Lock, M. 2002. Human body parts as therapeutic tools: Contradictory discourses and transformed subjectivities. *Qualitative Health Research* 12: 1406-18.

Luxemburg, R. 2003. *The accumulation of capital.* London, New York: Routledge.
Mol, A., and J. Law. 1994. Regions, networks and fluids: Anaemia and social topology. *Social Studies of Science* 24: 662-88.
Nagan, W. 2002. International intellectual property, access to health care and human rights: South Africa v. United States. *Florida Journal of International Law* 14: 155-91.
Ngwane, T. 2003. Sparks in the township. *New Left Review* 22(II): 37-56.
Offe, C. 1982. Some contradictions of the modern welfare state. *Critical Social Policy* 2: 17-45.
Pattinson, S.D. 2003. Paying living organ donors. *Web Journal of Current Legal Issues* 3. Available at: <http://webjcli.ncl.ac.uk/2003/issue3/pattinson3.html> (Last accessed: 13 April 2005).
Petras, J. 1999. Globalization: A critical analysis. *Journal of Contemporary Asia* 29: 3-37.
Pollock, A. 2004. *NHS plc.* London: Verso.
Price, D., A.M. Pollock, and J. Shaoul. 1999. How the World Trade Organization is shaping domestic policies in health Care. *The Lancet* 354: 1889-91.
Radcliffe Richards, J., A.S. Daar, R.D. Guttmann, R. Hoffenberg, I. Kennedy, M. Lock, R.A. Sells, and N. Tilney. 1998. The case for allowing kidney sales. *The Lancet* 352: 1950-2.
Richardson, E., and Turner, B.S. 2002. Bodies as property: From slavery to DNA Maps. In: *Body Lore and Laws,* eds. A. Bainham, S. Day-Sclater, and M. Richards, 29-42. Oxford: Hart.
Sassen, S. 1991. *The global city: New York, London, Tokyo.* Princeton: Princeton University Press.
Scheper-Hughes, N. 2000. The global traffic in human organs. *Current Anthropology* 41: 1-33.
———. 2001. Commodity fetishism in organs trafficking. *Body and Society* 7: 31-62.
———. 2003. Keeping an eye on the global traffic in human organs. *The Lancet* 361: 1645-8.
Sexton, S. 2001. *Trading health care away? GATS, public services and privatisation. Corner House Briefing No. 23.* London: The Corner House.
Shaffer, E., and J. Brenner. 2004. Trade and health care: Corporatizing vital human services. In *Sickness and Wealth* eds. M. Fort, M.A. Mercer, and O. Gish, 79-94. Cambridge (Mass): South End Press.
Sheldon, S. 1997. *Beyond control: Medical power and abortion law.* London: Pluto.
Singer, P. 2004. *One world: The ethics of globalization.* New Haven, London: Yale University Press.
Steering Committee on Bioethics and European Health Committee. 2003. *Replies to the questionnaire for members on organ trafficking.* Strasbourg: Council of Europe.
Stiglitz, J. 2002. *Globalization and its discontents.* Harmondsworth: Penguin.
Sutcliffe, B. 2002. How many capitalisms? Historical materialism in the debates about imperialism and globalization. In: *Historical materialism and globalization,* eds. M. Rupert, and H. Smith, 44-58. London, New York: Routledge.
Titmuss, R.M. 1970. *The gift relationship: From human blood to social policy.* London: George, Allen & Unwin.
UNCTAD, 2002. *World investment report.* New York: United Nations.
UNESCO, 1989. *Human rights aspects of traffic in body parts and human fetuses for research and/or therapeutic purposes.* Paris: UNESCO.
van der Mei, A.P. 2003. Cross-border access to health care within the European Union: Recent developments in law and policy. *European Journal of Health Law* 10: 369-80.
Wallace, T. 2003. NGO dilemmas: Trojan Horses for global neoliberalism? In: *The new imperial challenge: Socialist register 2004,* eds. L. Panitch, and C. Leys, 202-19. London: Merlin Press.
Wallerstein, I. 1984. *The politics of the world economy. The states, the movements and the civilizations.* Cambridge: Cambridge University Press.
Whitty, N. 1999. "In a perfect world": Feminism and resource allocation in health care. In: *Feminist perspectives on health care law,* eds. S. Sheldon, and M. Thomson, 135-53. London: Cavendish.
Wilkinson, S. 2003. *Bodies for sale. Ethics and exploitation in the human body trade.* London: Routledge.
Wood, E.M. 2002. *Empire of capital.* London: Verso.
World Bank. 2003. *World Development Report 2004.* Washington DC: World Bank.
World Health Organization. 1991. Human Organ Transplantation. A Report on the Developments under the Auspices of the WHO. *International Digest of Health Legislation.* Dordrecht: Martinus Nijhoff.
World Medical Association. 2000. Statement on human organ and tissue donation and transplantation. Geneva: WMA.

CHAPTER TWELVE

IAN FRECKELTON

HEALTH PRACTITIONER REGULATION

Emerging patterns and challenges for the age of globalization

A number of patterns in the regulation of health practitioners can be identified internationally since the early 1990s. This period of time has seen a significant homogenization and globalization in regulation but it has also seen the emergence of complex new trends and difficulties. Significant changes in "consumer culture" and the emergence of the "information age" have engendered new attitudes toward and expectations of health service provision and the regulation of practitioners.

A number of developments in medical regulation can be identified in the United Kingdom, Canada, Australia and New Zealand. This chapter will concentrate upon these four jurisdictions, omitting for the most part reference to the diverse systems of regulation that apply in the different states of the United States. It will argue that the increased mobility of the medical workforce and a preparedness on the part of the media to sensationalize doctors' misconduct has led to a firming up of both intranational and transnational registration and disciplinary processes.

A number of particular features can be identified internationally in relation to medical regulation. First, the dominance of the paradigm of self-regulation and peer review has significantly reduced with the removal of decision making in a number of jurisdictions from practitioners' regulatory bodies. Second, involvement of non-practitioners in the investigative and decision making processes in relation to complaints against health practitioners has significantly escalated. Third, regulation has started to extend into the provision of complementary services (Petersen, forthcoming; Freckelton 2000). Fourth, where issues of misconduct have been proved, and ongoing risk has been identified, the consequences for practitioners have become more adverse in an attempt to protect the community against practitioners who have engaged in misconduct or who are impaired in their competency or performance. Fifth, this has resulted in an increasing legalism in the disciplinary-conduct component of health practitioner regulation, manifested by initial hearings being more assertively contested and the incidence of appellate litigation from the first instance decisions of boards and tribunals rising. Sixth, there

B. Bennett & G.F. Tomossy (eds.), Globalization and Health: Challenges for Health Law and Bioethics,
187-206.
© 2006 *Springer. Printed in the Netherlands.*

has been a growing recognition of the potential for systems deficiencies, health issues and skills deficits to result in unsatisfactory outcomes in terms of practitioner conduct. An aspect of this recognition has been a tendency for regulation to focus more upon competencies and less on the need to impose strong disciplinary penalties to regulate health practitioners. Seventh, pressures are building internationally to ensure that practitioners are fit to practice in an ongoing way via revalidation schemes as a means of monitoring ongoing competence amongst health care practitioners.

However, other features have emerged along with the moves toward international consensus as to how errant practitioners should be regulated. Medical, dental and nursing practitioners, in particular, have become more mobile with a high volume of professionals moving from one country to another. An increasing trend is that the developed world experiences shortages in its locally trained practitioners and seeks to supplement this deficit by recruiting practitioners trained and initially registered in other countries. Often this means divesting third world countries of a percentage of their practitioners in order to meet the ever-increasing demand for health services in the first world. In turn this generates regulatory issues for accreditation of international medical graduates, for consistent levels of training, for the quality of service provision by practitioners culturally unfamiliar with their new environments, and for international sharing of information about the good standing of practitioners who may encounter regulatory difficulties in one country and then move elsewhere to recommence practice untrammeled by the obloquy that their behavior has generated.

At a broad level, globalization of health service provision regulation poses important questions about international work force movement and the potential for third world countries to be deprived of badly needed health care resources. Ironically, some countries have been so denuded of Western medical resources by departure of their medical practitioners that complementary medicine and quasi-trained practitioners of Western medicine have become the ascendant mode of health service provision. The call has been made by representatives of a number of developing countries (Mariba 2004) that the developed world should refrain from "poaching" scarce health care personnel from developing countries or, if they persist in doing so, should recompense developing countries.

This chapter explores the moves toward uniformity of approach amongst health regulatory bodies internationally. It also endeavors to grapple with the advantages and dilemmas posed by the globalization of health service delivery and regulation. Principally it discusses the regulation of medical practitioners but much of its content is applicable also to the regulation of health practitioners such as nurses, dentists, psychologists, physiotherapists, chiropractors, osteopaths, podiatrists and optometrists, as well as the regulation of complementary health care practitioners such as traditional Chinese doctors, naturopaths and herbalists.

GENERAL INFLUENCES ON THE REGULATORY ENVIRONMENT

Consumerism

The environment within which health services are delivered in developed countries changed fundamentally during the latter part of the twentieth century and the early years of the twenty-first century. Perhaps most significantly, the extent to which members of the general community were prepared to defer uncritically to the knowledge and judgment of medical and other health practitioners changed. The paternalism that had characterized health service provision became anachronistic and the fulcrum of power shifted from practitioners to patients. Condescension and arrogance by practitioners was vilified, examples of gross derelictions of care by practitioners were sensationally canvassed in the media, and the rights of patients to receive what they or the general community were paying for in terms of health care became a consumerist catch-cry. This was especially manifested in expectations by patients about decision making in relation to treatment provided to them.

To a significant extent in the public mind in Western countries health services became commodified. Lobbyists and health rights activists insisted that patients be referred to as "clients" or "consumers" and that practitioners be accountable in the same way as other providers of services to the community. To the distress of many medical practitioners especially (Psychiatrists Working Group 1997; Mendelson 2004), health service provision became regarded by many as a business and entitlements to good service from health practitioners were said to be analogous to rights in relation to local council services, social security benefits or after-sales service by department stores.

Sometimes services were received by patients as an incident of being a member of the community – via the National Health Service, Medicaid or Medicare. Sometimes patients paid for them – on occasions, as in relation to cosmetic surgery or laser eye vision remediation, they paid for them heavily. However, the consumerist assertion came to be that patients had a right to a duty of care on the part of health service providers and to be treated with respect, dignity and autonomy, enforceable by civil litigation, occasionally by criminal litigation and through the making of complaints to regulatory bodies, if the service provision failed to match expectations or was of an unacceptable quality. While such entitlements in fact generally matched the content of ethical codes binding health practitioners, what had changed was the assertion of entitlements made by reference to the codes, rather than commitment by practitioners to the values (Bloch and Coady 1997).

Information availability

Entitlements of patients to information so that they can make informed choices about their treatment was at the heart of the health care practitioner-patient revolution of the 1990s. An aspect of consumerist entitlements was the impact of the information age upon the dynamic between health practitioners and their patients.

With the abandonment of the "doctor knows best" value system came the capacity of many patients in the era of the Internet to have recourse to information about illnesses, treatments, complication rates (Duckett, Hunter and Rassaby 1999; Freckelton 1999) and alternative approaches to orthodox health service provision (Weir 2000). Often such information was not mediated by the tools which would enable effective evaluation of the vast amounts of information available electronically. However, the mystique of medicine and all of the health services, save perhaps some of the complementary areas of practice, seemed to lose their magic. The aura of inaccessibility which had made medicine an art more than a science was lost in a drive to evaluate evidence-based medicine, gauge success rates, develop doctors' report cards, examine risk potentials, and stipulate quality assurance levels within health care delivery.

The rhetoric of medicine and other areas of health service delivery became bureaucratized and management discourse, outcome objectives, key performance indicators and econometrics became fundamental to health policy development. This was coterminous with recognition by governments that the health dollar was likely to impact increasingly upon the viability of Western economies as developed countries' populations aged and the capacity for technological extension of lives became a serious drain on budgets.

Another aspect of the information age is a capacity on the part of investigative and monitoring agencies to assemble information about the patterns of health service provision by individual practitioners, comparing them against the mean and inquiring as to the reasons for departures from the statistically normal. Such concerns commenced in the 1990s in a number of countries to form the basis for both criminal prosecutions and regulatory hearings in relation to overservicing, inappropriate prescribing and self-prescribing. In Australia, Patient Services Review committee hearings are regularly convened to deal with medical practitioners suspected of having abused their state-granted entitlements under Medicare.

Two consequences attached to the phenomena accompanying the health information age. The first is a preparedness on the part of a percentage of patients to lodge complaints with regulatory bodies and even to initiate civil litigation when they do not receive the information to which they consider they were entitled before submitting to a particular form of treatment. The second is an inclination on the part of organs of government, such as those administering public health programs, monitoring practitioners' prescription patterns, especially in relation to narcotics and benzodiazepines, and overseeing the potential for excessive service provision also to initiate investigations into practitioners' conduct.

Public health perspectives

An element of regulation of health professionals views adverse outcomes and iatrogenic harm as a public health risk. It attempts to document the incidence and nature of the expression of grievances as a means of identifying poor service provision and thereby to protect the public from the phenomenon of poorly performing and dangerous practitioners. Such an approach analyzes patterns in the notification of complaints and classifies in a quasi-epidemiological way those factors which place members of the community at risk and which are over-represented amongst practitioners the subject of complaints, proven and not proven.

Thus, complaints are viewed as an opportunity to isolate patterns of poor practitioner behavior (Wilson 1999, 179; Paterson 2004) and characteristics which may be amenable to remediation or change; thereby complaints are seen as a form of health information that creates means of improving the quality of health service provision.

Therapeutic jurisprudence

Along with the emergence of regulatory hearings as a significant area of legal practice has come a growing reflectiveness by a number of regulatory bodies about whether their processes and outcomes are effective in protecting the public and whether they are functioning in a style that is in conformity with the principles of health disciplines. Part of this has been framed in terms of the therapeutic jurisprudence movement which seeks to study the role of the law as a therapeutic agent. It focuses on the law's impact on emotional life and on psychological well-being (Wexler and Winick 1996, xvii). Therapeutic jurisprudence scholars have argued that involvement in a legal process can act as a clinical intervention, which in turn can have positive repercussions for the safety of the community.

At its core, therapeutic jurisprudence offers a lens through which it is possible to analyze the content of the law, the interpretation of the law and legal processes in a sophisticated and sensitive way. In particular, it gives specific recognition to the health repercussions of each aspect of the operation of the law in practice (Wexler and Winick 1991; 1996; 2003; Winick 1997a; 1997b; Diesfeld and Freckelton 2003). It highlights the potential for investigations taking place in the shadow of the law and for legal hearings in both courts and tribunals to have counter-therapeutic as well as pro-therapeutic outcomes. Over the last one and a half decades, therapeutic jurisprudence has generated fresh perspectives on many components of the law for more attuned awareness of the mental health consequences of different practice options within the content and application of the law. In this regard, it draws upon victimology, public health law, critical legal studies, law and society, restorative justice, positive and pragmatic psychology and other modern perspectives on the law. It offers the possibility for the law to draw creatively for both its processes and its outcomes from the insights of the mental health professions.

In the context of investigations and hearings conducted by health regulatory bodies, it has been argued that a closer awareness of the dynamics involved in the lodgment of complaints, the experience of being a complainant and a person complained about, and of the factors that are likely to lead to improved performance and reduced potential for recidivism can raise the likelihood of therapeutic outcomes and reduce the likelihood of counter-therapeutic consequences to the management of complaints about health practitioners (Freckelton and Flynn 2004; Freckelton and

List 2004). This has both a practitioner orientation and a public health focus in terms of seeking to reduce the incidence of unsatisfactory conduct by registered health practitioners.

It is likely that this reconceptualized approach toward health practitioner regulation will erode some measure of the adversarialism emerging in regulatory hearings and result in an impetus to address the core issue of how most effectively the public can be protected against practitioner error, incompetence, and misconduct. It is consistent with the emerging concentration, described below, upon factors such as practitioner health, skills and competency as frequently responsible for the commission of unsatisfactory professional behavior.

EMERGING PATTERNS IN REGULATION

Self-regulation and peer review

The dominant reference point for regulation of professionals has traditionally been peer review, the notion that those best able to evaluate professional conduct are colleagues of the practitioner (Daniel 1990). It has been thought that peers are well-positioned to gauge propriety of conduct because of themselves being in a comparable, or the same, area of practice and thus able to assess whether a practitioner has fallen below what is to be expected of a practitioner of good repute and competency. In addition, peers should be prepared to make adverse findings so as to demarcate between ethical and unethical conduct and thereby to be seen to be ready to police high standards of behavior within their profession – by so doing they are upholding the integrity and standing of their profession.

However, critics have contended that instead a "brotherhood ethic" has led to a propensity on the part of peer review bodies to be unduly tolerant of unprofessional behavior and a pattern of adopting "tap on the shoulder" strategies for regulating unprofessional conduct rather than robust adverse findings at formal inquiries: sexual transgressions have been identified as a particular example of inappropriate leniency (Thomas 2004; Rogers 2004). Also it has been suggested (Victorian Department of Human Services 2003) that the quality of decision making by peers has lacked the hallmarks of transparency and accountability which should characterize the determination of serious allegations of misconduct against members of a profession.

Further, it has been said that members of the public will not have confidence in a system which consists of complaints against health practitioners being resolved by persons likely to know and to have worked with the practitioners the subject of complaint. Some disciplinary systems have incorporated investigative, prosecutorial and arbitral functions within the same body (Victorian Department of Human Services 2003). Particular concerns have been identified in relation to such forms of peer regulation on the basis of the extent to which complainants might fear that they would not receive a full and dispassionate treatment of their grievances.

The international trend appears to be away from unfettered self-regulation of health practitioners. An example of this, discussed below, is the increasing role of non-practitioner members of regulatory bodies. Another is the fact that in a number of jurisdictions a statutory measure of unprofessional conduct is conduct of a lesser standard than is reasonably to be expected of a practitioner by members of the public.[1] This explicitly incorporates into the criteria for unsatisfactory professional behavior an extraprofessional measure, one that is reflective of contemporary community standards.

Non-practitioner participation in regulation

It is noteworthy that lay participation on regulatory bodies is comparatively recent. Such a non-practitioner role reduces the extent to which decision making fully constitutes peer review but it provides an external perspective which reduces the force of the criticism that regulatory bodies are "old boys' clubs" "looking after their own." In Victoria the eleven regulatory bodies for health practitioners each have two lay participants out of between nine and twelve members. By contrast, the United Kingdom General Medical Council (GMC) has 40 percent lay membership and on performance assessments of medical practitioners there must always be a non-medical participant. In Canada there is generally significant lay participation on disciplinary hearings although lay involvement in performance assessments is not yet part of the process. The international trend appears to be toward a greater role for non-practitioner contribution to assessment of practitioners and formal decision making; conversely there is some attrition of the extent of peer review as the major paradigm.[2]

Regulation of complementary medicine

The fact that such a considerable percentage of health service provision is via practitioners of alternative or complementary medicine has brought with it the call for greater regulation of health care practices which are otherwise dependent upon the bringing of civil actions by aggrieved patients or the exclusion from professional associations of poorly performing practitioners (Bensoussan and Myers 1996; Petersen, forthcoming). An added fillip to extended regulatory arrangements has been the recognition that a percentage of complementary practitioners do not function as complementary practitioners – rather, they function separately and apart from orthodox medicine, do not cross-refer patients and may not be adequately aware of the potentiating or toxic effects of combined doses of complementary and orthodox medications (Freckelton 2003). This has been of particular concern in the context of treatments for cancer and other diseases which are potentially curable by orthodox medicine.

In Victoria a bold step has been taken in relation to traditional Chinese medicine and is mooted in relation to naturopathy. The *Chinese Medicine Registration Act 2000* (Vic) provided that the practice of Chinese medicine, defined to include any form of acupuncture or treatment with or dispensing of Chinese herbs, is illegal save by persons duly registered by the Chinese Medicine Registration Board. This has led to many difficult grandfathering decisions in relation to long-time practitioners with qualifications from tertiary courses other than those in Australia but has provided a fillip for Chinese medicine to function in a more evidence-based way and as a genuine complement to orthodox Western medicine. It has also encouraged practitioners of Chinese medicine to become better schooled in the effects and characteristics of orthodox pharmacotherapy as well as Chinese herbs, the latter of which also can have lethal toxicities.

It is likely that the recognition of the potentially dangerous nature of a number of the complementary therapies will lead to registration processes by formal bodies, rather than simply professional associations, principally as a mechanism for removing practitioners with inadequate qualifications and skill levels and also as a means of inducing practitioners to function in a complementary rather than competitive capacity.

Tougher disciplinary penalties

Significant changes can be identified in the forms of disposition meted out to health practitioners found to have engaged in significantly unprofessional conduct since the 1980s. Most prominent amongst the changes have been the sanctions imposed for sexual misconduct, conflict of interest and lack of probity. This has been matched by a concomitant shift in regulators' rhetoric. The significant change in the approach by courts and disciplinary tribunals has been characterized by an overt acknowledgment of the disempowerment generally existing between a patient and a health care practitioner, an imbalance prone to exploitation if the practitioner does not act ethically (Freckelton 1994). Robust critiques of abuse of power and predatory behavior by health professionals published in professional literature (Walton 2002; Gutheil and Gabbard 1993; Lucire 2002; Farber, Novack and O'Brien 1997; Gartrell et al. 1992; Searight and Campbell 1993) appear to have played a role in shifting the views of regulatory bodies. This has resulted in a pattern of firmer first instance decision making by regulatory bodies and tribunals in relation to crossing of sexual boundaries, especially by psychiatrists and psychologists.

The issue of the incompatibility of dishonesty and on-going registration of a health care practitioner has come before the health regulatory tribunals and the courts on a number of occasions. In the important decision of the Privy Council in *Patel v. General Medical Council*[3] a doctor appealed from a decision by the Professional Conduct Committee of the General Medical Council in England, ultimately to the Privy Council, against a decision that his name be erased from the practising register for various acts of dishonesty. The Council held that "For all professional persons including doctors a finding of dishonesty lies at the top end in the spectrum of gravity of misconduct."[4]

What can be discerned is a readier response by first instance regulators to pronounce decisively that a variety of forms of conduct on the part of health professionals constitute behavior that is prima facie incompatible with registered status. This compares with the more "understanding" response of earlier generations of regulators.

Increasing legalism

Matters involving serious allegations against health practitioners have the potential to result in deregistration, suspension or the imposition of conditions upon registration. It goes without saying that such sanctions can be personally and professionally devastating for practitioners.

It is apparent from the volume of appellate litigation in relation to health practitioners against whom adverse findings are made at first instance in the United Kingdom, Australia, New Zealand and Canada that the area of "disciplinary law" is expanding as a field of legal practice. There is an increasing tendency for technical points to be argued on behalf of practitioners, for legal principles to be explored and challenged and for decisions adverse to practitioners to be appealed. In Australia there is no doubt that first instance hearings are becoming longer and more complex. They are tending to involve more senior lawyers and to involve the taking of more technical points (Searles 1992). The decisions of the Professional Conduct Committees of the General Medical Council in the United Kingdom and in Victoria of the Medical Practitioners Board, the Psychologists Registration Board, the Dental Practice Board and the Chinese Medicine Registration Board and of the Civil and Administrative Tribunal on de novo review are posted on the Internet, enabling transparency of reasoning with the result that the shaming and the improprieties of practitioners are highly visible for their colleagues and even for current and potential patients.

The combination of factors has resulted in something of an impetus in the direction of adversarialism for matters that reach the formal decision making phase in Australia, New Zealand, the United Kingdom and Canada. In turn this has provided an incentive for regulatory authorities to reconfigure allegations of impropriety into matters of health, performance and competence in need of investigation, rather than to conduct inquiries which are likely to lead to heavily contested and stigmatizing hearings.

Awareness of broader causes for adverse outcomes

There is a growing recognition that there can be multiple causes for individual instances of unsatisfactory practitioner conduct. These include systems factors, health problems experienced by practitioners and deficiencies in skills and competence.

Systems issues

Irvine (1997a; 2004) has argued that there has been a problematic emphasis in the past on reactiveness to specific events, rather than the early recognition and proactive response to potential problems within medical service delivery (see also Cruess, Cruess and Johnston 2000). With some regularity, errors and unsatisfactory outcomes for which particular practitioners are responsible can be the product of complex institutional and systems factors. Many examples can be advanced. For instance, diminished concentration on the part of a registrar can be caused by his or her having worked a series of long shifts at the direction of hospital management. It accomplishes little to discipline the practitioner for an error brought about by exhaustion for which the practitioner bears little responsibility.

Likewise, hospital policy may require the discharge of certain categories of patients at an arbitrary time subsequent to surgery. The fact that a patient in the care of a surgeon or physician suffers complications which could have been identified and addressed had the patient continued to be an inpatient is technically the responsibility of the practitioner. However, resources issues which resulted in administrative decisions which in turn generated premature patient discharges may constitute the background within which a clinical error in judgment should be evaluated.

Such matters are ill-addressed by disciplinary hearings and have significant consequences for individual practitioners, palatable as such results may be for adversely affected patients or patients' families.

Health problems

A variety of factors can be responsible for unsatisfactory levels of performance by health practitioners. Numbered amongst these can be health conditions, ranging from cognitive impairment arising from age-related deterioration, brain injury or psychiatric illness to physical ailments, such as chronic pain or heart disease or other conditions to substance dependencies. A number of medical regulatory bodies in Australia, Canada and the United Kingdom have set up or facilitated health programs for doctors so as to maximize the prospects of practitioners receiving the assistance they need to address their health problems and thereby render them fit once more for practice.

Australian regulatory bodies tend to take a low key approach with health issues, eschewing formal hearings as often as possible, and encouraging practitioners to have breaks from work to address their health concerns and to return to practice subject to mutually agreed restrictions on practice for a period of time until medical reports suggest that the practitioner's condition has been adequately addressed. Thus, the approach is one of working with practitioners in order to facilitate rehabilitation and provision of education about the risks of not having a family doctor, working too hard, becoming too stressed and yielding to substance dependence.

Performance deficits

One of the notable features of regulation of health professionals is that cross-pollination amongst international bodies and legislatures is resulting in a reduction in the disciplinary orientation of regulation. Part of what is taking place is a redefinition of the culture of professionalism. For instance, Donald Irvine, president of the GMC, argued in 1997 that such a culture should be inclusive, involving all doctors and embracing continuing medical education, personal professional development, clinical audit and quality improvement methods. He contended that it should have six core components:

- A clear ethical framework and, where possible, the use of explicit professional and clinical standards;
- Effective local professional regulation for maintaining good practice;
- Regular publication by the Royal Colleges and others of data showing doctors' involvement in continuing medical education, audit, and other performance related activities;
- Sound local arrangements for recognizing dysfunctional doctors early and for taking appropriate action;
- Well defined criteria and pathways for referral to the GMC when severely dysfunctional doctors cannot or should not be managed locally;
- At all stages, practical help and support so that doctors who get into difficulties can be restored to full practice wherever possible.   (Irvine 1997b)

Many generic allegations are made against health practitioners, including: sexual misconduct, blurring of boundaries between practitioner and client, dishonesty, breach of confidentiality, conflict of interest, incompetence in the provision of treatment, conduct of procedures without suitable advice about options and potentially adverse consequences, and unsatisfactory communication generally. However, often particular instances of what might be classified as unprofessional conduct within these categories are indicative of deficits in skill or competence.

Starting with initiatives in Canada (Weymouth 1999; Hall et al. 1999; Page et al. 1995), a new approach has been trialed in jurisdictions in North America (Norma 1993), the United Kingdom (McCaul 2002),[5] Australia[6] and New Zealand (Competence Screening 2003), whereby in appropriate cases an investigation is directed not to whether a practitioner has engaged in unprofessional conduct but into the practitioner's professional competence and skill levels. Such an assessment is generally peer-dominated and conducted by reference to professional guidelines and practices, to evaluate whether the practitioner's performance is substandard by reference to accepted norms and procedures. If it is, the orientation of regulatory bodies is to address the deficits by mandated further education, counseling or supervision, or sometimes by restrictions on practice. As with the approach in relation to health-caused deficiencies of practice, the attempt is to draw upon the skills possessed by a practitioner and enhance the potential of the practitioner to practice safely and in accordance with professional standards. Where possible, this is done collaboratively by respected colleagues of the practitioner, at the aegis of the regulatory body, identifying deficits and setting out a roadmap toward remediation.

Weymouth (1999, 2) has described as the most common deficiencies identified in British Columbia insufficient medical records, lack of referral letters to consultants, inadequate sterilization of equipment and lack of an emergency kit (ibid.). He has commented that "Physicians with deficiencies have often developed habits characterized by inattention to details" (ibid.). Of the 1,434 primary or subsequent visits for performance assessment in British Columbia, 91.8 percent of physicians were found to have no significant deficiencies in their medical practice (ibid.).

In New Zealand between 1998 and April 2003, 314 doctors were referred to the Medical Council out of concern about their performance. Doctors with New Zealand qualifications were the least likely to have concerns expressed about them – only 2.6 percent of doctors registered in New Zealand. Of the 314 expressions of concern, 105 became actual competence assessments, 39 percent of those reviews demonstrating deficiencies in practice. The areas reviewed were clinical (62), communication (24), prescribing (17), surgical (16), records (12), boundaries (4) and other (8) (Medical Council News 2003, 3-5).

The emerging focus on performance assessment and monitoring builds upon the trend toward mandatory Continuing Professional Development (CPD) (Peck 2000; Davis et al. 2003; Parboosingh 2003; World Federation for Medical Education 2003; Department of Human Services 2003) and potential re-accreditation on the basis of demonstrated ongoing competency (Trunkey and Botney 2001). In the United Kingdom, which leads the world in this regard, a precondition to ongoing registration is proof of participation in CPD and accreditation. This is a dramatic shift away from the notion that mere possession of tertiary qualifications is sufficient for a lifetime license to practice. It provides a means whereby the general community can have a measure of confidence that health practitioners are staying abreast of current developments, are not isolated from their colleagues and remain well regarded and competent from the informed perspective of their professional peers. A corollary of this approach, though, is that fewer matters are likely to be dealt with as instances of unprofessional conduct, the focus increasingly being not upon individual acts that are unsatisfactory but upon what is giving rise to substandard performance and behavior likely to be indicative of defective skills, technique or knowledge.

However, the days are still relatively early in relation to performance-based resolution of complaints and notifications about practitioner performance. While the application of such an approach to the skills and competence of doctors and dentists is immediately plausible – at least in some areas of practice, such as anesthesia and endodontics – the application is less clear in relation to general practice and psychiatry amongst doctors and a number of other areas of practice by other health practitioners, such as psychology and nursing because of the logistical difficulties of assessment of substantial levels of practice. The boundary lines between conduct and performance also have to be worked through, so that it is clear whether a conduct or performance investigation will be triggered in given circumstances. In addition, the cost and delay ramifications of performance-based investigation have to be resolved as peer assessment tends to be time-consuming and costly – who is to pay for such investigations? There is also the question of how acceptable such an approach is to aggrieved patients who may have suffered adverse consequences to what they perceive as practitioner misconduct. While many complainants cite as their reason for lodging complaints their hope that the practitioner will be prevented from behaving toward others as they have toward them, the behind-the-scenes mentoring and guidance directed toward returning the practitioner to practice may not provide a high level of satisfaction to complainants.

Moves toward revalidation

A significant development, likely to be implemented incrementally internationally in varying forms (Rogers 2004; Dauphinee 1999; Southgate and Dauphinee 1998; Norcini 1999; Newble, Paget and McLaren 1999) is the proposal that doctors in the United Kingdom will in due course need a license to practice (Cunningham and Norman 2000; Southgate et al. 2001; McKinley, Fraser and Baker 2001). To retain their license, doctors will have to "revalidate," by demonstrating, at regular intervals to the GMC that they remain up-to-date and fit to practice. Henceforth revalidation and licensing will constitute the cornerstone of medical practitioners' accountability to United Kingdom patients and the wider public.[7] This is not a wholly new phenomenon, clinical governance previously being regarded as based on principles of local oversight and accountability – formal procedures were introduced into the National Health System from April 1999 which included annual appraisal for doctors, based on the guidance given by the GMC's *Good Medical Practice*.

Historically, doctors were admitted to the register on qualification (or on assessment, if they qualified outside the United Kingdom). No further checks were done unless the doctor's performance gave rise to concerns that led to a referral to the fitness to practice procedures (Cunningham et al. 1997). The change is that revalidation will shift the emphasis of regulation away from qualifications alone towards a regular assessment of whether the doctor remains up-to-date and fit to practice (McKinley, Fraser and Baker 2001). It will require all doctors to establish that they reflect meaningfully on their practice, using information gathered through audit and in other ways, and to seek the views of others on their performance, throughout their medical careers.

Every doctor who holds a license to practice will be required to revalidate – normally once every five years. If a doctor does not participate co-operatively in revalidation, their registration will be at risk of being withdrawn. Doctors will be asked to show, by the production of verifiable evidence, that their practice over the period specified has been in line with the principles set out in *Good Medical Practice*. All doctors will have to provide a description of their medical practice, including clinical and non-clinical activities, throughout the specified period.

Doctors will be obliged to collect and retain data and information routinely which is drawn from their day-to-day medical practice to support their revalidation ("the doctor's folder"). The expectation is that most practitioners will already have been doing this for a variety of reasons, including reporting into local clinical governance systems and preparing for annual appraisals.

The revalidation process is intended also to ensure that doctors regularly reflect on their practice. For most doctors, the vehicle for delivering that regular reflection will be an annual appraisal but if a practitioner cannot participate in such an appraisal scheme, the GMC will call for the doctor's folder. The GMC will also require verifiable evidence that there are no significant local concerns about a doctor's practice. Suitable evidence, for a doctor who works in a managed environment, will be confirmation by a person with formal clinical governance responsibilities that there are no unresolved concerns about the doctor's practice. In the absence of this, the results of validated questionnaires, completed by professional colleagues and, where appropriate, patients also, may constitute acceptable evidence. If a doctor cannot produce an appropriate clinical governance certificate or results from validated questionnaires, the GMC will request to see the doctor's folder. Finally, doctors will be required to provide evidence as to their good health and probity. All practitioners will be required to provide a personal declaration, certified by another doctor who is registered with the GMC.

What is being attempted in the United Kingdom therefore is an extensive, monitored, mandated regime for ensuring medical practitioners' fitness to practice. It is open to criticism for its bureaucratization, its dependence upon what are likely to be pro forma processes and the burden that it will impose upon practitioners to compile relevant documentation – more time on paperwork and less on practice. However, its aims to weed out practitioners who are lacking in core competencies, who are out-of-date, who are ill thought of by peers or whose health is declining will be politically attractive well beyond the United Kingdom.

Health practitioner mobility

Another phenomenon which has become notable is the growing mobility of medical practitioners from one country to another. Many Western countries, the United Kingdom, Canada, Australia and New Zealand, for instance, have become heavily dependent upon overseas trained doctors to maintain needed levels of health service delivery (Birrell and Hawthorne 2004). What is taking place with the aging of Western countries' populations is an increasing demand for medical services, a demand that is not being met by the availability of local doctors, especially in rural areas.

In Australia during the late 1990s the Commonwealth Government financially assisted new state-based rural recruitment agencies whose task it was to procure medical practitioners for country areas which were struggling to fill vacancies for general practitioners. The number of visas issued to overseas trained doctors for the temporary entry visa category (422) increased from 664 in 1993-1994 to 2,496 in 2002-2003 (Birrell and Hawthorne 2004).

A disproportionate percentage of such practitioners come from Africa, in particular Nigeria and South Africa, a number of Asian countries, such as India,

Sri Lanka and Malaysia, and in more latter times a variety of countries in eastern Europe. This raises a series of complex issues for regulation. The big picture issue is that the developed world is denuding a number of developing countries of their expensively trained medical practitioners (Mariba 2004). Individual doctors have the potential to earn much greater incomes in Western countries and to work in more up-market circumstances. Understandably, they are tempted to exercise their entitlement to take advantage of such opportunities for their own professional development and the financial security of themselves and their families.

However, their own countries have often invested significant sums in their education, and are receiving little return on their investment. The result is a medical brain drain and the fact that medical resources are reaching unsustainable levels in some doctors' countries of origin. This is prompting considerable resentment and calls for preclusion periods on "doctor harvesting" or compensation to be paid by Western countries availing themselves of doctors qualified in developing countries (Mariba 2004).

There are other issues posed by medical mobility also. There is the potential for practitioners who have been the subject of adverse determinations by regulatory authorities or about whom it is known that they have engaged in unprofessional conduct to move from one jurisdiction to another without their background being known. The facility for moving countries can be a license to engage in serially improper conduct if regulatory authorities do not adequately communicate with each other. A change that has occurred over the past decade is that such authorities increasingly effectively co-operate to communicate information about doctors and other health practitioners who move from one country to another. However, there are many countries where regulation is relatively unstructured and not particularly effective. In respect of such countries there are ongoing risks posed by medical migration.

The increasing demand for medical practitioners qualified in other countries poses its own problems in relation to assessment of the sufficiency of tertiary qualifications obtained and specialist accreditation obtained in some countries. This generates pressure for internationally consistent medical courses, accreditation criteria and preconditions for membership of learned colleges in respect of specialist practitioners. In the Australian context, Birrell and Hawthorne have identified that over 3,000 overseas trained doctors are currently working in Australia, by contrast with a population of between 20,000 and 25,000 permanent non-specialist medical practitioners. They argue that as a result of these changes in medical manpower:

> an increasing minority of [overseas trained doctors] may not have experienced a training program equivalent to that prescribed for local doctors. Very few have had to undergo rigorous examination of their skills prior to practice commencement. An increasing proportion is coming from nations where the training programs are not tailored to the health profile characteristics of Australian patients. When such doctors have been required to undergo the Australian Medical Council accreditation examinations, the proportion succeeding has been modest.   (Birrell and Hawthorne 2004, 97)

The difficulties exist in many parts of the world. A 2004 study recounted, for example, that the Centre for Spinal Injuries in Boxburg, near Johannesburg in South Africa, was the referral centre for the whole region. On the same day in 2000 its two anesthetists were recruited by a Canadian institution opening a new Spinal Injuries Unit. A consequence of the loss of these two key staff members was the temporary closure of the unit (Martineau, Decker and Bundred 2004; Dobson 2004, 419). The same study reported that India has lost up to £2.7 billion in investment in the training of doctors since 1951 and that Ghana had lost around $US60 million. The authors maintained that there was a "carousel" movement of doctors, the Canadian provinces of Alberta and Saskatchewan, in particular, having been recruiting for general practitioners to work in remote rural areas. They identified the same phenomenon in respect of South Africa in its attempts to fill rural posts (Martineau, Decker and Bundred 2004).

This raises an issue faced in a number of countries – the challenge for regulating doctors who find themselves in culturally different surroundings, perhaps in a rural environment with many expectations that they had previously not had to meet. Such practitioners may encounter prejudices and milieus for which they are ill-prepared and environments into which they do not readily meld. This can be as much the case socially as professionally. The combination can result in substandard or culturally inappropriate behaviors which in turn generate grievances and discomfort from members of the public. Sometimes the causes can be simple ignorance, sometimes they are culturally attributable, and sometimes they are the product of stress-induced poor decision making.

In Australia a response in 2003-2004 was an increased vetting of overseas trained applicants for medical registration – not just to assure facility with the English language but also to raise the likelihood that practitioners have a reasonable awareness of the kinds of practice which they are anticipated to enter (Hawthorne, Birrell and Young 2003). In addition, improvements were made in acculturation of practitioners and provision of mentoring and supports. It is too early to evaluate whether these measures have improved the experience of newly arrived doctors and in turn whether this has enhanced their levels of performance and reduced allegations of misconduct.

THE FUTURE

Globalization of medical practice has resulted in the development amongst first world countries of more demanding and increasingly consistent expectations about what constitutes ethical practice and how departures from ethical practice should be dealt with. Recent measures adopted, for instance, by the United Kingdom GMC, have tended to be emulated in other jurisdictions. Examples include the increased lay involvement in regulation of medical practitioners and a growing focus upon performance and competency evaluation rather than response to conduct infractions. Another example is likely to be an insistence for the purposes of registration upon compliance with significant ongoing professional development.

With such changes have come both a new level of censoriousness in Western countries about breaches of ethics involving relationships with patients, lack of probity, breaches of confidentiality, provision of treatment without adequate consent, and conflicts of interest. In turn, the harsher sanctions meted out for such improprieties have raised the stakes of disciplinary proceedings, resulting in growing legalism. This has flowed on to a higher profile for regulatory authorities and tribunals, escalating registration fees and a new level of media scrutiny of accountability for decision making and mechanisms in relation to health practitioners.

The scrutiny has commenced to raise issues about differential levels of skill and training amongst practitioners from different countries. With the escalation in and impetus toward medical migration from developing countries to first world countries has come a welter of "big picture" ethical issues and the potential for anomalies in terms of values and behaviors on the part of practitioners moving from one cultural context to another. This is requiring increased communication amongst regulators, accrediting agencies and tertiary institutions. The result is likely to be a level of homogenization of approach amongst countries but a risk that first world countries will disproportionately profit from the existence of practitioners seeking to enjoy the fruits of their training from developing countries by practice in countries where the financial rewards are more lucrative. This aspect of medical migration is shaping as a significant international issue, once more requiring co-operation and consistency, in order to ensure that patients are protected against incompetent and unethical behavior by doctors whose dangerous conduct transcends national boundaries.

NOTES

1. See, for example, *Medical Practice Act* 1994 (Vic), section 3.
2. In Victoria, there are *de novo* rehearings by the Victorian Civil and Administrative Tribunal from decisions of the Medical Practitioners Board. Often they are conducted by a legal member of the tribunal sitting alone. In 2004 this was the subject of criticism in Freckelton and Flynn (2004, 100).
3. *Patel v. General Medical Council* [2003] UKPC 16.
4. Ibid. at [10].
5. See the work undertaken by the National Clinical Assessment Authority. However, the General Medical Council performance procedures rely upon the parameters of the GMC's third edition in 2001 of *Good Medical Practice*.
6. *Medical Practice Act* 1994 (Vic), Part 3, Division 3.
7. See: <http://www.gmc-uk.org/revalidation/index.htm> (Last accessed: 31 March 2005).

REFERENCES

Bensoussan, A., and S.P. Myers. 1996. *Towards a safer choice: The practice of traditional Chinese medicine in Australia.* MacArthur: University of Western Sydney.

Birrell, B., and L. Hawthorne. 2004. Medicare Plus and overseas trained medical doctors. *People and Place* 12(2): 84-100.

Bloch, S., and M. Coady. 2000. *Codes of ethics for the professions.* Melbourne: Melbourne University Press.

Competence Screening – the State of Play. 2003. *Medical Council News (NZ)* 35: 3.

Cruess, R.L., S.L. Cruess and S.E. Johnston. 2000. Professionalism: An ideal to be sustained. *Lancet* 356: 156-9.

Cunningham, J.P.W., and G.R. Norman. 2000. Certification and recertification: Are they the same? *Academic Medicine* 75(6): 617-9.

Cunningham, J.P.W., E. Hanna, J. Turnbull, T. Kaigas, and G.R. Norman. 1997. Defensible assessment of the competency of the practicing physician. *Academic Medicine* 72(1): 9-12.

Daniel, A. 1990. *Medicine and the state: Professional autonomy and public accountability.* Sydney: Allen & Unwin.

Dauphinee, W.D. 1999. Revalidation of doctors in Canada. *British Medical Journal.* 319: 1188-90.

Davis, D., M. Evans, A. Jadad, L. Perrier, D. Rath, D. Ryan, G. Sibbald, S. Straus, S. Rappolt, M. Wowk, and M. Zwarenstein. 2003. The case for knowledge translation: Shortening the journey from evidence to effect. *British Medical Journal.* 327: 33-5.

Diesfeld, K., and I. Freckelton, eds. 2003. *Involuntary detention and therapeutic jurisprudence: International perspectives on civil commitment.* Dartmouth: Ashgate.

Dobson, R. 2004. Poor countries need to tackle the brain drain. *British Medical Journal* 329: 419.

Duckett, A., L. Hunter, and A. Rassaby. 1999. *Health services policy review: Discussion paper.* Melbourne: Victorian Government Department of Human Services.

Farber, N.J., D.H. Novack, and M.K. O'Brien. 1997. Love, boundaries and the patient-physician relationship. *Archives of Internal Medicine* 157: 2291-4.

Freckelton, I. 1994. The sexually exploitative doctor. *Journal of Law and Medicine* 1: 203-4.

———. 1999. Materiality of risk and proficiency assessment: The onset of the era of healthcare report cards? *Journal of Law and Medicine* 6(4): 313-8.

———. 2000. Regulation of Chinese medicine. *Journal of Law and Medicine* 8: 5-17.

———. 2003. Shakoor v. Situ. *Journal of Law and Medicine* 10: 268-70.

———. 2004. Regulating forensic deviance: The ethical responsibilities of expert report writers and witnesses. *Journal of Law and Medicine.* 12: 141-9.

Freckelton, I., and J. Flynn. 2004. Paths toward reclamation: Therapeutic jurisprudence and the regulation of medical practitioners. *Journal of Law and Medicine* 12(2): 91-102.

Freckelton, I., and D. List. 2004. The transformation of regulation of psychologists by therapeutic jurisprudence. *Psychiatry, Psychology and Law* 11(2): 296-307.

Gartrell, N.K., N. Milliken, W.K. Goodson III, S. Thiemann, and B. Lo. 1992. Physician-patient sexual contact: Prevalence and problems. *Western Journal of Medicine* 157(2): 139-43.

General Medical Council, United Kingdom. Undated. *Licensing and Revalidation.* Available at: <http://www.gmc-uk.org/revalidation/index.htm> (Last accessed: 12 January 2005).

———. 2001. *Good Medical Practice.* Available at: <http://www.gmc-uk.org/standards/good.htm> (Last Accessed: 20 January 2005).

Gutheil, T.G., and G.O. Gabbard. 1993. The concept of boundaries in clinical practice: Theoretical and risk-management dimensions. *American Journal of Psychiatry* 150(2): 188-96.

Hall, W., C. Violato, R. Lewkonia, J. Lockyer, H. Fidler, J. Toews, P. Jennett, M. Donoff, and D. Moores. 1999. Assessment of physician performance in Alberta: The physician achievement review. *Canadian Medical Association Journal* 161(1): 52-7.

Hawthorne, L., B. Birrell, and D. Young. 2003. *The retention of overseas trained doctors in general practice in regional Victoria.* Melbourne: Rural Workforce Agency Victoria.

Irvine, D.H. 1997a. The performance of doctors I: Professionalism and regulation in a changing world. *British Medical Journal* 315: 1540-2.

———. 1997b. The performance of doctors II: Maintaining good practice, protecting patients from poor performance. *British Medical Journal* 314: 1613-5.

———. 2004. Time for hard decisions on patient-centered professionalism. *Australian Medical Journal* 181(5): 271-4.

Lucire, Y. 2002. Sex and the practitioner: The victim. *Australian Journal of Forensic Science* 34: 17-24.

Mariba, T. 2004. Challenges in international medical regulation. Paper read at Australian and New Zealand Medical Boards and Councils Combined Annual Conference, Towards 2010: Doctors and the community, 16 October 2004, Melbourne.

Martineau, T., K. Decker, and P. Bundred. 2004. Brain drain of health professionals: From rhetoric to responsibility. *Health Policy* 70: 1-10.

McCaul, J. 2002. The Scottish approach to poorly performing doctors. Available at: <http://careerfocus.bmjjournals.com/chi/content/full/324/7334/S51?ct> (Last accessed: 11 January 2005).

McKinley, R., R.C. Fraser, and R. Baker. 2001. Model for directly assessing and improving clinical competence and performance in revalidation of clinicians. *British Medical Journal* 322: 712-5.

Mendelson, D. 2004. HealthConnect and the duty of care: A dilemma for medical practitioners. *Journal of Law and Medicine* 12: 69-79.

Newble, D., N. Paget, and B. McLaren. 1999. Revalidation in Australia and New Zealand: Approach of Royal Australasian College of Physicians. *British Medical Journal* 319: 1185-8.

Norcini, J.J. 1999. Recertification in the United States. *British Medical Journal* 319: 1183-5.

Norman, G.R., D.A. Davis, S. Lamb, E. Hanna, P. Caulford, and T. Kaigas. 1993. Competency assessment of primary care physicians as part of a peer review program. *Journal of the American Medical Association* 270(9): 1046-51.

Page, G. G., J. Bates, S.M. Dyer, D.R. Vincent, G. Bordage, A. Jacques, A. Sindon, T. Kaigas, G. R. Norman, and M. Kopelow. 1995. Physician-assessment and physician-enhancement programs in Canada. *Canadian Medical Association Journal* 153(12): 1723-8.

Parboosingh, J. 2003. CPD and maintenance of certification in the Royal College of Physicians and Surgeons of Canada. *The Obstetrician and Gynaecologist* 5: 43-9.

Paterson, R. 2004. Complaints and quality: Handle with care! *New Zealand Medical Journal.* 117(1198): 970-3.

Peck, C., M. McCall, B. McClaren and T. Rotem. 2000. Continuing medical education and continuing professional development: International comparisons. *British Medical Journal* 320: 432-5.

Petersen, K. Forthcoming. Regulation of complementary health practitioners. In: *Disputes and dilemmas in health law*, eds. I. Freckelton, and K. Petersen. Sydney: Federation Press.

Psychiatrists Working Group. 1997. *She won't be right, mate! The impact of managed care on Australian psychiatry and the Australian community.* Melbourne: Psychiatrists Working Group.

Rogers, S. 2004. Culling bad apples: Blowing whistles and the *Health Practitioners Competence Assurance Act* 2003 (NZ). *Journal of Law and Medicine* 12: 119-133.

Searles, D. 1992. Professional misconduct – unprofessional conduct. Is there a difference? *Queensland Law Society Journal* 22(3): 239-44.

Searight, H.R., and D.C. Campbell. 1993. Physician-patient sexual contact: Ethical and legal issues and clinical guidelines. *Journal of Family Practice* 36(6): 647-53.

Southgate, L. and D. Dauphinee. 1998. Maintaining standards in British and Canadian medicine: The developing role of the regulatory body. *British Medical Journal* 316: 697-700.

Southgate, L., J. Cox, T. David, D. Hatch, A. Howes, N. Johnson, B. Jolly, E. Macdonald, P. McAvoy, P. McCrorie, and J. Turner. 2001. The assessment of poorly performing doctors: The development of the assessment programmes for the General Medical Council's performance procedures. *Medical Education* 35 (Suppl 1): 2-8.

Thomas, D. 2004. The co-regulation of medical discipline: Challenging medical peer review. *Journal of Law and Medicine* 11: 382-389.

Trunkey, D.D., and R. Botney. 2001. Assessing competency: A tale of two professions. *Journal of the American College of Surgeons* 192(3): 385-95.

Walton, M. 2002. Sex and the practitioner: The predator. *Australian Journal of Forensic Sciences* 34: 7-15.

Weir, W. 2000. *Complementary medicine: Ethics and law.* Brisbane: Prometheus Publications.

Wexler, D.B., and B.J. Winick. 1996. *Law in therapeutic key: Developments in therapeutic jurisprudence.* Durham: Carolina Academic Press.

———, eds. 1991. *Essays in therapeutic jurisprudence.* Durham: Carolina Academic Press.

Weymouth, V. 1999. Office medical practice peer review: the British Columbia experience. Paper presented at the Parlons Qualite Conference, 17 June 1992, Montreal. Revised October 1999.

Winick, B.J. 1997a. *The right to refuse mental health treatment.* Washington DC: American Psychological Association.

————. 1997b. *Therapeutic jurisprudence applied: Essays on mental health law*. Durham: Carolina Academic Press.

Winick, B.J., and D.B. Wexler. 2003. *Judging in a therapeutic key*. Durham: Carolina Academic Press.

Wilson, B. 1999. Health disputes: A window of opportunity to improve health services. In *Controversies in health law*, ed I. Freckelton, and K. Petersen, 179-92. Sydney: Federation Press.

World Federation for Medical Education. 2003. Continuing Professional Development of Medical Doctors WFME Global Standards for Quality Improvement. Result from Task Force Seminar, 25-27 October 2002.

INDEX

207

International Library of Ethics, Law, and the New Medicine

International Library of Ethics, Law, and the New Medicine

24. L. Romanucci-Ross and L.R. Tancredi: *When Law and Medicine Meet: A Cultural View*. 2004
ISBN 1-4020-2756-7

25. G.P. Smith II: *The Christian Religion and Biotechnology*. A Search for Principled Decision-making. 2005 ISBN 1-4020-3146-7

26. C. Viafora (ed.): *Clinical Bioethics*. A Search for the Foundations. 2005 ISBN 1-4020-3592-6

27. B. Bennett and G.F. Tomossy (eds.): *Globalization and Health*. Challenges for health law and bioethics. 2005 ISBN 1-4020-4195-0

28. C. Rehmann-Sutter, M. Düwell and D. Mieth (eds.): *Bioethics in Cultural Contexts*. Reflections on Methods and Finitude. 2006 ISBN 1-4020-4240-X

29. S.E. Sytsma: *Ethics and Intersex*. 2006 ISBN 1-4020-4313-9

30. M. Betta (ed.): *The Moral, Social, and Commercial Imperatives of Genetic Testing and Screening*. The Australian Case. 2006 ISBN 1-4020-4618-9